for that matter smelled. (Yes, the Internet has a smell, Blum is here to report.) Though less ethereal and a bit dingier, the Internet that Blum's beautifully lucid prose makes real turns out to be if anything a more marvelous place than the cloudy dreamland we'd imagined."   —Donovan Hohn, author of *Moby-Duck: The True Story of 28,800 Bath Toys Lost at Sea and of the Beachcombers, Oceanographers, Environmentalists, and Fools, Including the Author, Who Went in Search of Them*

"We think of the Internet as a kind of ether, a magical way of transporting words and images from anywhere to anywhere else. But there is a vast physical infrastructure behind all that magic, and in *Tubes*, Andrew Blum, one of our best writers on the built environment, discovers it and turns it into a compelling story of an altogether new realm where the virtual world meets the physical."

—Paul Goldberger, author of *Why Architecture Matters*

"Andrew Blum's journey in search of the Internet—the actual physical wires-and-tubes reality of our twenty-first-century ethereal life—takes him from the subterranean world under New York's manhole covers to Cornish fields and Portuguese cable-laying vessels. At once funny, prosaic, sinister, and wise, Blum's tale is a beautifully written account of the true human cost of all our remote connectivity."

—Bella Bathurst, author of *The Lighthouse Stevensons*

TUBES

ALSO BY ANDREW BLUM

The Weather Machine

TUBES

A JOURNEY

TO THE

CENTER

OF THE

INTERNET

ANDREW BLUM

An Imprint of HarperCollinsPublishers

HarperCollins books may be purchased for educational, business, or sales promotional use. For information, please email the Special Markets Department at SPSales@harpercollins.com.

A hardcover edition of this book was published in 2012 by Ecco, an imprint of HarperCollins Publishers.

FIRST ECCO PAPERBACK EDITION PUBLISHED 2013.
SECOND ECCO PAPERBACK EDITION PUBLISHED 2019.

Designed by Leah Carlson-Stanisic

Library of Congress Cataloging-in-Publication Data has been applied for.

ISBN 978-0-06-285030-0

19 20 21 22 23   LSC   10 9 8 7 6 5 4 3 2 1

For Davina and Phoebe

It is not down in any map; true places never are.

—HERMAN MELVILLE

Somehow I knew that the notional space behind all of the computer screens would be one single universe.

—WILLIAM GIBSON

Contents

Introduction

In the process of writing this book, I visited dozens of corporate data centers, submarine cable landing stations, and Internet exchange points. I crossed through "man traps" (double doors that unlock one at a time, for security), peered into "cages" (fenced-in equipment enclosures), and traced the wires snaking through "ladder racks" (ceiling-mounted cable shelves). Again and again I toured these places with helpful guides, pointing at boxes with blinking lights and asking, "What's this?" Their answers were often thorough and generous. Never in those tours did anyone come up short, slamming closed a half-open door or shunting me past some secret installation.

I wouldn't go so far as to say that this physical infrastructure of the Internet was literally open to everyone—much of it was locked away inside unmarked buildings behind high fences. But its ethos was open. The Internet is a network of networks; it demands interdependence and cooperation, and a willingness on the part of its operators to worry not only about their own networks, but their interconnections with other networks. Infrastructure, like architecture, reflects the ideals of its creators, and the Internet's astonishing emergence in the 1990s and 2000s depended on this collective spirt. It was vivid in my reporting: the network engineers, data center builders, and undersea cable technicians were welcoming and uncensored, genuinely eager for me to understand their piece of the Internet and, once permission was granted, to come inside their discreet and secure facilities, ready to show off every nook and cranny, or, in a few cases, leave me alone to have the run of the place. Their stories are in this book.

But in the years since, that ethos has been threatened. It is easy to pinpoint when I was startled by these changes. Just after 5 p.m. on June 5, 2013, the *Washington Post* published the first of what became a long series of bombshells based on the documents leaked by National Security Agency contractor Edward Snowden. I was on a train from Washington, DC, clattering across the Maryland suburbs, not far from the NSA's headquarters, when the article dropped. It described a program codenamed "PRISM," which gave US and British intelligence agencies access to the user data held by nine major American Internet companies—Google, Facebook, Apple, and Microsoft among them. Thanks to both spy movies and past press reports, I had a rudimentary understanding of the broad capabilities of the NSA to intercept phone calls and emails, and a loose faith in the appropriateness of those actions. But I knew immediately that this revelation was of an entirely different magnitude—this story would break through the chatter, with consequences both for our understanding of the Internet's infrastructure, and the role of Silicon Valley's giant companies in our lives.

With PRISM, the line between the government's collection of data for espionage and Silicon Valley's collection of data for profit wasn't so much crossed as erased. As journalists Laura Poitras and Barton Gellman wrote, "There has never been a Google or Facebook before, and it is unlikely that there are richer troves of valuable intelligence than the ones in Silicon Valley." This collection of data was subject to the approval of a top-secret intelligence court (and was immediately denied by the companies involved) but it nonetheless made it clear that the terms of the debate had permanently shifted, and the conventional wisdom about Silicon Valley's benevolence was about to swing wildly. All this happened long before the 2016 election, before uncountable privacy breaches (both accidental and deliberate), and before an endless series of missteps that continuously exposed an inveterate carelessness with keeping our private information private. Before 2013, I had noticed a tendency to give the giants of Silicon Valley the benefit of the doubt—to put what they do for us (whether free email, or web searches, or storing our documents) ahead of what they extract from us. After Snowden, and more so each year, that is no longer the case.

The special knowledge I had from studying the physicality of the Internet made those first Snowden revelations particularly startling. I knew how much these companies knew about us, but—more important—I also knew how little we knew about them. I saw how deliberately Facebook and Google, among others, kept the specifics of their work hidden, making it nearly impossible to understand even the simplest details of how their sites operated. (To take one basic example, why Facebook's News Feed puts some posts ahead of others remains inscrutable.) I had experienced their obfuscations firsthand, heard their non-answers or, worse, endured their condescending reassurances that they had the whole privacy and security situation under control. The PRISM story was bigger. It suggested that there were whole continents of things I didn't even know I didn't know. Along with the further revelations of tapping and data collection gleaned from the Snowden documents, it began to seem as if there were a shadow Internet, built to siphon the contents of the main Internet. This was hard to imagine. I knew the effort and expense of keeping the first Internet up and running; the existence of a sort of second, secret, Internet was an astonishing revelation. But more to the point, if there was a parallel Internet, how had I missed it? And what else had I missed?

In retrospect there were clues. Most notably, I remember an interview with the CEO of a trans-Atlantic undersea cable, who helpfully answered all my questions. But when I asked about visiting his cable landing station he dismissed the possibility, which was unusual. "In summary we don't really welcome it, as part of our security is just being low profile," he said. When the Snowden documents trickled out, I saw that one of the key tapping locations of the GCHQ (the United Kingdom's version of the NSA) was nearby. Had he demurred about a visit because I might point at a box, ask "What's this?" and thereby expose a tap that it otherwise required a monumental leak to reveal? That seemed unlikely. But his reticence (which I'd hardly registered at the time) only highlighted all that I hadn't seen—not merely the so-called dark web, with its hackers and illegal marketplaces, but the full breadth of information being collected by the Internet's biggest players, like Google, Amazon, and Facebook.

The physical infrastructure of the Internet I describe in this book is still there. Its structure and organizing principles are mostly the same as when I began my reporting. What has changed is the valence of it all: the stakes are higher, the darkness more exposed. As I reflect on the evolution of the Internet's infrastructure over the past decade, both what we know about it and why it matters, what unfortunately becomes clear is that our worst fears about the Internet have come true. The network has crept further and further into our private lives, with webcams and digital assistants, smart watches and thermostats, apps for taxis, apps for pizza, and apps to track our health. In the process, it has become more inscrutable. Each new revelation of tracking, surveillance, stolen passwords, or algorithmic biases only clarifies the need to understand the systems Silicon Valley has created. That is challenging—especially when there's a strong business case for keeping us ignorant. But I know that understanding the Internet's physical pieces is a step toward better understanding the more complex virtual ones: the algorithms and code upon which so much of our lives depend.

In all this, there has been one major change for the better: parts of the Internet's infrastructure are more visible than before. Most notably, soon after the publication of this book, Google loosened its policy of secrecy when it came to its data centers—among the Internet's most prominent repositories. Google welcomed a journalist (*Wired*'s Steven Levy) inside a data center for the first time and released a set of photographs (which Internet sleuths quickly determined had been photoshopped; old habits die hard, apparently). I can't know for sure, but I'd like to claim some credit for that new attitude. Perhaps it had just become better for business to share the details of its infrastructure, rather than to hide them? Either way, Google's insistence that its users didn't care about that infrastructure, and just wanted its email to work, was obviously no longer viable.

When I visited Google's data center in The Dalles, Oregon, in 2011, the only sign outside read "Voldemort Industries." The land had been purchased through an anonymous LLC. City officials were cautious about even mentioning the company's name. Today, Google touts each new data center or undersea cable investment with a blog post and ribbon-cutting ceremony. To my greatest surprise, some of the buildings that once went sign-less have now become canvasses for a well-publicized mural project, to decorate their blank walls. My favorite among the artworks was conceived by Jenny Odell, who used fragments of images from Google Maps as the basis for large geometric forms hand-painted on the side of a massive Google data center in Oklahoma. Her artwork's conceit is that these images passed through the building, thereby telling us something about what goes on inside. Her leap from the bits inside to the paint outside is abstract, but I find the symbolism of it thrilling, and the work itself beautiful—a perfectly pitched public-facing art project for what was only recently a building whose existence Google denied. If the old Voldemort sign was the worst kind of inside joke, making light of Google's omniscience and secrecy, Odell's mural is at least an effort at communicating the human uses of the building.

It has been thrilling to see the geeky and overlooked spaces of the Internet brought into the light and acknowledged as the key underpinning of our culture. I laughed when the HBO sitcom *Silicon Valley* used the inside of a data center—and the often-eccentric characters to be found there—as a running gag. But I nearly spit out my coffee when I first saw the images from Chanel's Spring/Summer 2017 fashion show, held in Paris in the busy fall of 2016. The runway was decked out as a data center, with bundles of multicolored wires and blinking lights, their colors picked up in the design of the clothes. "The data center is something of our time," Karl Lagerfeld, the fashion house's legendary director, said. It was his response to the dominance of Instagram and social media, and an effort to bring the physicality of the Internet into the physicality of a fashion show. Culturally at least, the Internet's infrastructure was hidden no more, and that thrilled me. On the evening of May 18, 2017, a more cheerful tidbit crossed my screen. Katy Perry posted a tweet to her one hundred million followers: "Is the Internet real? Can we physically go there?"

Yes, Katy, yes we can. And going there, to the Internet, is a good first step out of the dark forest which we find ourselves lost in today.

March 2019
Brooklyn, New York

Prologue

On a bitterly cold day a few winters ago, the Internet stopped working. Not the whole Internet, only the section that resides in a dusty clump beside my living room couch. There's a black cable modem with five green lights, a blue telephone adapter the size of a hardcover book, and a white wireless router with a single illuminated eye. On good days they all blink happily at one another, satisfied with the signals coming through the wall. But on that day their blinking was labored. Web pages loaded in fits and starts, and my phone—of the "voice over IP" variety, which sends calls over the Internet—made everyone sound like a scuba diver. If there were little men inside these boxes, then it was as if they had suddenly become prone to naps. The switch itself had fallen asleep.

The repairman arrived the next morning, full of assurances. He attached an electronic whistle—it looked like a penlight—to the living room end of the cable, and then began to trace its path, searching for clues. I followed him, first outside to the street, then down into the basement and through a hatch to the backyard. A rusty switch box was caught in a web of black cables and bolted to a brick wall. Disconnecting them one at a time, he screwed a tiny speaker into each one until he found the one that whistled: audible proof of a continuous path between here and there.

Then his eyes lifted ominously to the sky. A squirrel scampered along a wire toward a battleship gray enclosure affixed like a birdhouse to a pole. Anemic urban vines wrapped around it. Animals chew on the rubber coating, the repairman explained. Short of rewiring the whole backyard there was nothing he could do. "But it might get better on its own," he said, and it did. But the crude physicality of the situation astonished me. Here was the Internet, the most powerful information network ever conceived! Capable of instantaneous communication with anyplace on earth! Instigator of revolutions! Constant companion, messenger of love, fountain of riches and beloved distraction. Stymied by the buckteeth of a Brooklyn squirrel.

I like gadgets. I will happily discuss the Internet as a culture and a medium. My mother-in-law calls me for tech support. But I confess that the substance of the thing—a "thing" that squirrels can nibble at—had escaped me. I may have been plugged in, but the tangible realities of the plug were a mystery to me. The green lights on the box in my living room signaled that "the Internet"—a singular unnuanced whole—was, to put it simply, *on*. I was connected, yes; but connected to what? I'd read a few articles about big factory-sized data centers filled with hard drives, invariably someplace far away. I'd unplugged and plugged back in my share of broken cable modems behind the couch. But beyond that, my map of the Internet was blank—as blank as the Ocean Sea was to Columbus.

That disconnect, if I can use that word, startled me. The Internet is the single biggest technological construction of our daily existence. It is vivid and alive on the screens all around us, as boisterous as a bustling human city. Two billion people use the Internet, in some form, every day. Yet physically speaking, it is utterly disembodied, a featureless expanse: all ether, no net. In the F. Scott Fitzgerald story "My Lost City," the protagonist climbs to the top of the Empire State Building and recognizes, crestfallen, that his city had limits. "And with the awful realization that New York was a city after all and not a universe, the whole shining edifice that he had reared in his imagination came crashing to the ground." I realized that my Internet had limits too. Yet, oddly, they weren't abstract limits but physical ones. My Internet was in pieces—literally. It had parts and places. It was even more like a city than I had thought.

The squirrel outage was annoying, but the sudden appearance of the Internet's texture was thrilling. I've always been acutely attuned to my immediate surroundings, to the world around me. I tend to remember places the way a musician does tunes or a chef, flavors. It's not merely that I like to travel (although I do), but more that the physical world is a source of constant, sometimes overwhelming, preoccupation. I have a strong "sense of place," as some people describe it. I like to notice the widths of the sidewalks in cities and the quality of light in different latitudes. My memories are almost always keyed to specific places. As a writer, that's often led me to the subject of architecture, but it's never been the buildings themselves that interest me most, rather the places the buildings create—the sum total of construction, culture, and memory; the world we inhabit.

But the Internet has always been a necessary exception to this habit, a special case. Sitting at my desk in front of a computer screen all day, and then getting up at the end of the day and habitually looking at the other, smaller screen I carry in my pocket, I accepted that the world inside them was distinct from the sensory world all around me—as if the screens' glass were not transparent but opaque, a solid border between dimensions. To be online was to be disembodied, reduced to eyes and fingertips. There wasn't much to do about it. There was the virtual world and the physical world, cyberspace and real places, and never the two shall meet.

But as if in a fairy tale, the squirrel cracked open the door to a previously invisible realm behind the screen, a world of wires and the spaces in between. The chewed cable suggested that there could be a way of stitching the Internet and the real world together again into a single place. What if the Internet wasn't an invisible elsewhere, but actually a *somewhere*? Because this much I knew: the wire in the backyard led to another wire, and another behind that—beyond to a whole world of wires. The Internet wasn't actually a *cloud;* only a willful delusion could convince anyone of that. Nor was it substantially wire*less*. The Internet couldn't just be everywhere. But then where was it? If I followed the wire, where would it lead? What would that place look like? Who would I find? Why were they there? I decided to visit the Internet.

Prologue

When in 2006 Senator Ted Stevens of Alaska described the Internet as "a series of tubes," it was easy to ridicule him. He seemed hopelessly, foolishly trapped in the old way of knowing the world, while the rest of us had skipped merrily into the future. Worse, he was supposed to know better. As chair of the US Senate's Commerce, Science, and Transportation Committee, Stevens had oversight for the telecommunications industry. But there he was behind the lectern of the Hart Building on Capitol Hill, explaining that "the Internet is not something that you just dump something on. It's not a big truck, it's a series of tubes, and if you don't understand those tubes can be filled, and if they're filled when you put your message in and it gets in line it's going to be delayed—by anyone that puts into that tube enormous amounts of material . . . *Enormous* amounts of material!" The *New York Times* fretted over the senator's cluelessness. Late-night comics showed side-by-side pictures of dump trucks and steel tubes. DJs mixed mash-ups of his speech. I made fun of him to my wife.

Yet I have now spent the better part of two years on the trail of the Internet's physical infrastructure, following that wire from the backyard. I have confirmed with my own eyes that the Internet is many things, in many places. But one thing it most certainly is, nearly everywhere, is, in fact, a series of tubes. There are tubes beneath the ocean that connect London and New York. Tubes that connect Google and Facebook. There are buildings filled with tubes, and hundreds of thousands of miles of roads and railroad tracks, beside which lie buried tubes. Everything you do online travels through a tube. Inside those tubes (by and large) are glass fibers. Inside those fibers is light. Encoded in that light is, increasingly, *us*.

I suppose that all sounds improbable and mysterious. When the Internet first took off, in the mid-1990s, we tended to think of it as a specific kind of place, like a village. But since then those old geographic metaphors have fallen out of favor. We no longer visit "cyberspace" (except to wage war). All the "information superhighway" signs have been taken down. Instead, we think of the Internet as a silky web in which every place is equally accessible to every other place. Our connections online are instantaneous and complete—except when they're not. A website might be "down" or our home connection might be wonky, but it's rare that you can't get to one part of the Internet from another—so rare that the Internet doesn't appear to have any parts at all.

The preferred image of the Internet is instead a sort of nebulous electronic solar system, a cosmic "cloud." I have a shelf filled with books about the Internet and they all have nearly the same picture on the cover: a blob of softly glowing lines of light, as mysterious as the Milky Way—or the human brain. Indeed, thinking of the Internet as a physical thing has fallen so far out of fashion that we're more likely to view it as an extension of our own minds than a machine. "The cyborg future is here," proclaimed the technology writer Clive Thompson in 2007. "Almost without noticing it, we've outsourced important peripheral brain functions to the silicon around us."

I know what that feels like, but I'm left wondering about all that "silicon around us." Clearly Thompson means our computers and smartphones and e-readers and whatever other devices we hold at arm's length. But it must also include the network be-hind them—and where's that? I'd feel better about outsourcing my life to machines if I could at least know where they were, who controls them, and who put them there. From climate change to food shortages to trash to poverty, the great global scourges of modern life are always made worse by not knowing. Yet we treat the Internet as if it were a fantasy.

The Silicon Valley philosopher Kevin Kelly, faced with this chasm between the physical here and the missing virtual there, became curious if there might be a way to think of them together again. On his blog he solicited hand sketches of the "maps people have in their minds when they enter the Internet." The goal of this "Internet Mapping Project," as he described it, was to at-tempt to create a "folk cartography" that "might be useful for some semiotician or anthropologist." Sure enough, one stepped forward out of the ether two days later—a psychologist and pro-fessor of media at the University of Buenos Aires named Mara Vanina Osés. She analyzed more than fifty of the drawings Kelly collected to create a taxonomy of the ways people imagined the Internet: as a mesh, a ring, or a star; as a cloud or a radial like the sun; with themselves at the center, on the bottom, the right or the left. These mental maps mostly divide into two camps: cha-otic expressions of a spidery infinity, like Jackson Pollock paint-ings; or an image of the Internet-as-village, drawn like a town in a children's book. They are perceptive, revealing plenty of self-consciousness about the way we live on the network. What strikes me, though, is that in no case do the machines of the In-ternet actually appear. "All that silicon" is nowhere to be found. We seem to have exchanged thousands of years of mental cartography, a collective ordering of the earth going back to Homer, for a smooth, placeless world. The network's physical reality is less than real—it's irrelevant. What Kelly's folk cartography portrayed most vividly was that the Internet is a landscape of the mind.

This book chronicles my effort at turning that imagined place into a real one. It is an account of the physical world. The Internet may seem to be everywhere—and in many ways it is—but it is also very clearly in some places more than others. The single whole is an illusion. The Internet has crossroads and superhighways, large monuments and quiet chapels. Our everyday experience of the Internet obscures that geography, flattening it and speeding it up beyond any recognition. To counter that, and to see the Internet as a coherent physical place all its own, I've had to tinker with my conventional picture of the world. At times this book's attention oscillates between a single machine and an entire continent, and at other times I simultaneously consider the tiny nano scale of optical switches and the global scale of transoceanic cables. I often engage with the most minute of timetables, acknowledging that an online journey of milliseconds contains multitudes. But it is a journey nonetheless.

This is a book about real places on the map: their sounds and smells, their storied pasts, their physical details, and the people who live there. To stitch together two halves of a broken world—to put the physical and the virtual back in the same place—I've stopped looking at web "sites" and "addresses" and instead sought out real sites and addresses, and the humming machines they house. I've stepped away from my keyboard, and with it the mirror-world of Google, Wikipedia, and blogs, and boarded planes and trains. I've driven on empty stretches of highway and to the edges of continents. In visiting the Internet, I've tried to strip away my individual experience of it—as that thing manifest on the screen—to reveal its underlying mass. My search for "the Internet" has therefore been a search for reality, or really a specific breed of reality: the hard truths of geography.

The Internet has a seemingly infinite number of edges, but a shockingly small number of centers. At its surface, this book recounts my journey to those centers, to the Internet's most important places. I visited those giant data warehouses, but many other types of places as well: the labyrinthine digital agoras where networks meet, the undersea cables that connect continents, and the signal-haunted buildings where glass fibers fill copper tubes built for the telegraph. Unless you're one of the small tribe of network engineers who often served as my guides, this is certainly not the Internet you know. But it is most definitely the Internet you use. If you have received an email or loaded a web page already today—indeed, if you are receiving an email or loading a web page (or a book) *right now*—I can guarantee that you are touching these very real places. I can admit that the Internet is a strange landscape, but I insist that it is a landscape nonetheless—a "netscape," I'd call it, if that word weren't already taken. For all the breathless talk of the supreme placelessness of our new digital age, when you pull back the curtain, the networks of the Internet are as fixed in real, physical places as any railroad or telephone system ever was.

In basest terms, the Internet is made of pulses of light. Those pulses might seem miraculous, but they're not magic. They are produced by powerful lasers contained in steel boxes housed (predominantly) in unmarked buildings. The lasers exist. The boxes exist. The buildings exist. The Internet *exists*—it has a physical reality, an essential infrastructure, a "hard bottom," as Henry David Thoreau said of Walden Pond. In undertaking this journey and writing this book, I've tried to wash away the technological alluvium of contemporary life in order to see—fresh in the sunlight—the physical essence of our digital world.

1

The Map

On the January day I arrived in Milwaukee, it was so cold that the streets themselves had blanched white. The city was born in 1846 out of three competing settlements at the edge of a broad harbor on the western shore of Lake Michigan. Four years after its founding, the Milwaukee & Waukesha Railroad linked the lake with the hinterland, and the rich wheat fields of the Midwest with the growing populations of the east. Before long, Milwaukeeans weren't only moving materials but processing them, making beer from hops, leather from cows, and flour from wheat. With the growing success of this industry—and the help of an influx of German immigrants—those first processing plants encouraged the growth of a broad range of precision manufacturing. The heart of the activity was the Menomonee Valley, a miasmic swamp that was steadily filled in to accommodate what was soon a coal-choked industrial powerhouse. "Industri-ally, Milwaukee is known across the face of the earth," the 1941 *WPA Guide to Wisconsin* rhapsodized. "Out of the city's vast machine shops come products that range from turbines weigh-ing 1,200,000 pounds to parts so minute as to be assembled only with the aid of magnifying glasses. Milwaukee steam shovels dug the Panama Canal; Milwaukee turbines harnessed Niagara Falls; Milwaukee tractors are in the fields of most of the world's agri-cultural regions; herring-bone gears made in Milwaukee oper-ate mines in Africa and Mexico, sugar mills in South America, and rolling mills in Japan, India, and Australia." Milwaukee had become the center of a far-reaching industrial colossus—known everywhere as "the machine shop of the world."

It didn't last forever. After World War II, the fixed steel lines of the railroads gave way to the more flexible movement of rub-ber tires over new roads. The hard networks became softer. And the Menomonee Valley started a steady decline, paralleling that of the nation's manufacturing more broadly. The United States became a country that produced ideas more than things. The "machine shop of the world" became the buckle of the Rust Belt. Milwaukee's factories were left abandoned—and then, only more recently, turned into condominiums.

But Milwaukee's industry didn't disappear entirely. It qui-etly holds on today, having moved out of the city and into the suburbs, like so much of American urban life. Early one morn-ing I followed its path, driving from a downtown hotel on a de-serted street to a new industrial neighborhood in the northwest corner of the city. I passed a McDonald's, a Denny's, an Olive Garden, and an IHOP, then took a left at a Honda dealer. High-tension power lines loomed overhead, and I bumped across a railroad spur that led the dozen miles back to the Menomonee Valley. Along a series of smooth, wide suburban streets was a concentration of industry that would have made William Harley and Arthur Davidson proud. In one building, they made beer cans; another, ball bearings. There were factories for car keys, airplane parts, structural steel, resistors, carbon brushes, mascot costumes, and industrial signs—that said things like WHEEL CHOCKS REQUIRED FOR LOADING AND UNLOADING. My destination was the tidy tan building across the road, with the giant "KN" painted on the side.

Kubin-Nicholson got its start in 1926, silk-screening movie posters from a print shop on Milwaukee's South First Street. In time, it branched out to signs for butchers, grocers, and department stores, before focusing on tobacco ads, printed in Milwaukee and pasted on billboards across the entire Midwest. Kubin-Nicholson was the "printers of the humongous." Its current press—as big as a school bus—sat within a cavernous hall. Its installation had taken a team of German engineers four months, flying home every other weekend to see their families. It was a rare beast, with fewer than twenty like it across the United States. And, on that morning, a frustratingly silent one.

The black ink was on the fritz. A call had been placed to the tech support people in Europe, who were able to log into the machine remotely to try to diagnose the problem. I watched from inside a glass-walled customer lounge, as the pressman peered into its innards, a cordless phone wedged in the crook of his neck, a long screwdriver in his hand. Beside me was Markus Krisetya, who had flown in from Washington to supervise the job on the press that day. He wanted to make sure the ink was precisely calibrated, so that just the right quantity of each color was distributed across the poster-sized paper. It wasn't the kind of thing that could be done over email. No digital scan would properly capture the nuance. FedEx would be far too slow for the back-and-forth, trial and error, required for the final settings. Krisetya accepted it as one of those things that still had to be done in person, a fact made even more surprising by what was being printed: a map of the Internet.

Krisetya was its cartographer. Each year, his colleagues at TeleGeography, a Washington, DC–based market research firm, polled telecommunications companies around the world for the latest information about the capacity of their data lines, their busiest routes, and their plans for expansion. TeleGeography's cartographers don't use any fancy algorithms or proprietary data analysis software. They worked an old-fashioned process of calling industry contacts and gaining their trust, then choosing just the right moment to make a few leaps of conjecture. Most of that effort goes toward a big annual report known as *Global Internet Geography,* or *GIG,* sold to the telecommunications industry for $5,495 a pop. But some of the key pieces of data are shunted into a series of maps of Krisetya's creation. One diagrammed the Internet's backbone architecture, the key links between cities. Another illustrated the quantities of network traffic, boiling trillions of moving bits down to a series of thick and thin lines. A third—the map on the press that morning in Milwaukee—showed the world's undersea communications cables, the physical connections between continents. All were representations of the spaces in between, the strands of connection that we typically ignore. The countries and continents were afterthoughts; their action was in the emptiness of the oceans. Yet these maps were also representations of physical things: actual cables, filled with strands of glass, themselves filled with light—amazing human constructions, of the kind a Milwaukeean would be proud.

Krisetya paid homage with his own sense of craft. When each map design was complete, he electronically transferred the file here to Milwaukee, then followed it himself. He'd stay at whatever downtown business hotel had a special, then head out here first thing in the morning, bringing nothing but a small gym bag, and his eyes. He knew big machines like this one. After college in the United States, he returned to his native Indonesia to work as a database systems engineer, mostly for the mining industry. Young, slight, with an easy manner, happy to fit in anywhere, eager for adventure, he'd show up at a remote encampment deep in the jungles, ready to tinker with their mainframes. As a boy, he'd drawn fantastical maps of *Dungeons & Dragons* realms, cribbed from bootlegged photocopied versions of the rulebooks that had somehow made their way to his home city of Salatiga. "I loved drawing stories on paper, and referencing distance in that strange manner," he told me, looking out at the silent press. "That's what got my fascination with maps started." It was only when he returned to the United States to study international relations in graduate school that his future wife, a geography student, encouraged him to take a cartography class taught by Mark Monmonier, author of the cult favorite *How to Lie with Maps*. The sly joke of the title is that maps never just show places; they express and reinforce interests. When TeleGeography offered Krisetya a job in 1999, he already knew the question: Maps project an image of the world—but what did that mean for the Internet?

With help from the tech support people in Germany, the pressman finally coaxed the giant machine to life, and its vibrations shook the door frames—*un-cha, un-cha, un-cha.* "I hear paper!" Krisetya cheered. A test print had been lain out on a large easel lit with klieg lights, like an operating table. Krisetya pulled off his thick-framed glasses and placed a magnifying loupe to his eye. I stood just over his shoulder, squinting at the bright lights, struggling to take in the world this map portrayed.

It was a Mercator projection, with the continents drawn in heavy black and the international boundaries etched, like afterthoughts, by thin scores. Rigid red and yellow lines striped the Atlantic and Pacific, jagged around the southern continents, and converged in key places: north and south of New York City, in the southwest of England, the straits near Taiwan, and the Red Sea—so tightly there that they formed a single thick mark. Each line represented a single cable, mere inches in diameter but thousands of miles in length. If you lifted one up from the ocean floor and sliced it crosswise, you'd find a hard plastic jacket surrounding an inner core of steel-encased strands of glass, each the width of a human hair and glowing faintly with red light. On the map it looked huge; on the ocean floor it would be a garden hose beneath the drifting sediment. It seemed to collapse the electronic global village upon the magnetic globe itself.

Krisetya examined every inch of the test print, pointing out imperfections. The pressman responded by moving levers up and down on a huge control panel, like the soundboard at a rock concert. Every few minutes, the giant press would spool up and spit out a few copies of the newest version. Krisetya would then go back over it again, inch by inch until finally, he put down his magnifier and nodded quietly. The pressman affixed a neon orange sticker to the map, and Krisetya signed it with a black marker, like an artist. This was the gold master, the definitive and original representation of the earth's underwater telecommunications landscape, circa 2010.

The networked world claims to be frictionless—to allow for things to be anywhere. Transferring the map's electronic file to Milwaukee was as effortless as sending an email. Yet the map itself wasn't a JPEG, PDF, or scalable Google map, but something fixed and lasting—printed on a synthetic paper called Yupo, updated once a year, sold for $250, packaged in cardboard tubes, and shipped around the world. TeleGeography's map of the physical infrastructure of the Internet was itself of the physical world. It may have represented the Internet, but inevitably it came from somewhere—specifically, North Eighty-Seventh Street in Milwaukee, a place that knew a little something about how the world was made.

To go in search of the physical Internet was to go in search of the gaps between the fluid and fixed. To ask, what could happen *anywhere*? And, what had to happen *here*? I didn't know this at the time, but in one of many strange ironies involved in visiting the Internet, over the next year and a half I would see TeleGeography's maps hanging on the walls of Internet buildings around the world—in Miami, Amsterdam, Lisbon, London, and elsewhere. Wedged into their plastic office-supply store frames, they were fixtures of those places, as much a part of the atmosphere as the brown cardboard shipping boxes piled up in the corners, or the surveillance cameras poking out from the walls. The maps were themselves like the dyes that trace fluid dynamics, their mere presence highlighting the currents and eddies of the physical Internet.

When the squirrel chewed through the wire in my backyard in Brooklyn, I had only the slightest inkling of how the Internet all fit together. I assumed my cable company must have a central hub somewhere—maybe out on Long Island, where its corporate headquarters was? But after that I could only imagine that the paths went everywhere, the bits scattering like Ping-Pong balls bouncing through dozens if not hundreds of tubes—more than could be counted, which was basically the same as saying none at all. I'd heard about an Internet "backbone," but the details were sketchy, and if it were truly a big deal, I figured I would have heard more. At the least, it would have occasionally become clogged or broken, bought or sold. As for international links, the undersea cables seemed mythic, like something out of Jules Verne. The Internet—other than as it appeared on my ever-present screen—was more conceptual than actual. The only concrete piece I had a clear image of were those big data centers, photographs of which I'd seen in magazines. They always looked the same: linoleum floors, thick bundles of cables, and blinking lights. The power of the images came not from their individuality, but from their uniformity. They implied an infinity of other machines standing invisibly behind them. As I understood it (but mostly didn't), those were the parts of the Internet. So what was I looking for?

I became an armchair traveler, querying network engineers with the same set of questions: How did the network fit together? What should I see? Where should I go? I started working up an itinerary, a list of cities and countries, of monuments and centers. But in the process I quickly stumbled on a more fundamental question about the network of networks: What was a network, anyway? I had one at home. Verizon had one too. So did banks, schools, and pretty much everyone else, some reaching across buildings, others across cities, and a few across the entire world. Sitting at my desk, I thought they all seemed to coexist, in relative peace and prosperity. Out there in the world, how did they all physically fit together?

Once I got my nerve up to ask the question at all, the whole thing started to make more sense. It turns out that the Internet has a kind of depth. Multiple networks run through the same wires, even though they are owned and operated by independent organizations—perhaps a university and a telephone carrier, say, or a telephone carrier contracted to a university. The networks *carry* networks. One company might own the actual fiber-optic cables, while another operates the light signals pulsing over that fiber, and a third owns (or more likely rents) the bandwidth encoded in that light. China Telecom, for example, operates a robust North American network—not as a result of driving bulldozers across the continent, but by leasing strands of existing fiber, or even just wavelengths of light within a shared fiber.

This geographic and physical overlapping was crucial to understanding where and what the Internet was. But it meant I had to get over the old, and really misleading, metaphor of the "information highway." It wasn't really that the network is a "highway" busy with "cars" carrying data. I had to acknowledge the extra layer of ownership in there: the network is more like the trucks on a highway than the highway itself. That allows for the likelihood that many individual networks—"autonomous systems," in Internet parlance—run over the same wires, their information-laden electrons or photons jostling across the countryside, like packs of eighteen-wheelers on the highway.

In that case, the networks that compose the Internet could be imagined as existing in three overlapping realms: logically, meaning the magical and (for most of us) opaque way the electronic signals travel; physically, meaning the machines and wires those signals run through; and geographically, meaning the places those signals reach. The logical realm inevitably requires quite a lot of specialized knowledge to get at; most of us leave that to the coders and engineers. But the second two realms—the physical and geographic—are fully a part of our familiar world. They are accessible to the senses. But they are mostly hidden from view. In fact, trying to see them disturbed the way I imagined the interstices of the physical and electronic worlds.

It was striking to me that I had no trouble thinking of a physical network of something, like a railroad or a city; after all, it shares the physical world in which we exist as humans, and which we learned as children to navigate. Similarly, anyone who spends time using a computer is at least comfortable with the idea of the "logical" world, even if we don't often call it that. We sign in to our home or office networks, to an email service, bank, or social network—logical networks all, which encompass our attention for hours on end. Yet we can't for the life of us grasp that narrow seam between the physical and the logical.

Here was the rarely acknowledged chasm in our understanding of the world—a sort of twenty-first-century original sin. The Internet is everywhere; the Internet is nowhere. But indubitably, as invisible as the logical might seem, its physical counterpart is always there.

I wasn't prepared for what that meant on the ground. Photographs of the Internet were always close-ups. There was no context, no neighborhood, no history. The places seemed interchangeable. I understood there were these layers, but it wasn't clear to me how they would appear in front of my face. The logical distinctions were, by definition, invisible. So then what was I going to see? And what was I really looking for?

A few days before I left for Milwaukee, I was emailing with a network engineer who'd been helping me with the basics of how the Internet fit together. He was a Wisconsin native, as it turned out. "If you're going to be in Milwaukee anyway, there is one spot you *must* hit," he wrote. There was an old building downtown "chock full of Internet." And he knew a guy who could show me around. "Have you seen *Goonies*?" he asked. "Bring your nice camera." After approving the test prints at Kubin-Nicholson, Krisetya usually spent the afternoon at the art museum before catching a flight home. But he was eager to come along. So we headed downtown to meet a stranger in a sandwich shop who was supposed to show us Milwaukee's Internet.

On his website, Jon Auer listed among his favorite books *Router Security Strategies* and *How to Win Friends & Influence People*. His Flickr page consisted mostly of photos of telecommunications equipment. In person, he had pink cheeks and metal-rimmed glasses, and on that frigid Wisconsin winter day he wore a hooded sweatshirt with no coat, and he carried a camouflage-patterned messenger bag. He fit the stereotype of a geek, but whatever social liability that might once have been, it had transformed into unadorned passion—and yielded a good job, running the network of a company that provides Internet access to towns across southeastern Wisconsin, mostly places too distant or too sleepy to attract the interest of the big telephone and cable operators. At lunch, he spoke almost in a whisper, conveying the impression that what we were about to do was slightly illicit, but not to worry. This was his turf, his backwoods. He had all the keys—and where he didn't, he knew the combination to the locks. He wrapped up his sandwich and led us out the back door of the shop, directly into the lobby of the building that turned out to be the center of Milwaukee's Internet.

Built in 1901 by a prominent Milwaukee businessman and once home to the Milwaukee Athletic Club, this building's days as a prestigious address were clearly long over. If in recent years the city had succeeded in revitalizing its downtown, that liveliness did not extend to this sad place. A sleepy-eyed guard sat listlessly behind a worn-out desk in the empty lobby. Auer nodded in her direction and led us down a narrow tiled passageway to the basement. Fluorescent lights buzzed dimly. There were dusty stacks of file boxes and precarious heaps of abandoned office furniture. The ceiling was totally obscured by a tangle of pipes and wires, twisted around one another like mangrove roots. They came in all sizes: wide steel conduits the diameter of dinner plates, orange plastic ducts like vacuum cleaner hoses, and the occasional single dangling black thread—the hackwork of a rushed network engineer. Auer shook his head at it, disapprovingly. I was struck with a more mundane thought: *look at all those tubes!* Inside of them were fiber-optic cables, glass strands with information encoded in pulses of light. In one direction, they went through the foundation wall and underneath the street, heading toward the highway—mostly to Chicago, Auer said. In the other direction, they crossed the basement ceiling to an old utility chase and upstairs to the offices-turned-equipment rooms of the dozen or so Internet companies that had colonized the building, feeding first off this fiber, and then off one another, one attracting the next, steadily displacing the cut-rate law firms and yellowed dentists' offices. Some were Internet service providers, like Auer's, that connected people in the surrounding area; others operated small data centers, which hosted the websites of local businesses on hard drives upstairs. Auer pointed out a steel box tucked into a dark corner, its LED lights blinking away. This was the main access point for Milwaukee's municipal data network, connecting libraries, schools, and government offices. Without it, thousands of civil servants would bang their computer mice against the desk in frustration. "All this talk about Homeland Security, but look what someone could do in here with a chainsaw," Auer said. Krisetya and I snapped pictures, the camera flashes blowing out the basement's dark crevices. We were spelunkers in a cave of wires.

Upstairs, the empty hallways smelled of mildew. We passed vacant offices, their doors cracked open. Auer's space looked like it belonged to a private eye in a film noir. The three small rooms had linoleum floors and worn-out Venetian blinds. The double-hung windows were thrown wide open to the winter, the cheapest way to keep the machines cool. The only evidence of the building's former opulence was a remnant scrap of mosaic floor tile, shattered in a corner like a broken mug. Auer's piece of the Internet was set unceremoniously on a raised platform: two man-sized steel racks, filled with a half-dozen machines, snug in a nest of cables. The key piece of equipment was a black Cisco 6500 Series router, the size of a few stacked pizza boxes, its chassis tattooed with bar-coded inventory labels and poked through by blinking green LEDs.

For the twenty-five thousand customers who relied on Auer's company to connect to "the Internet," this machine was the on-ramp. Its job was to read the destination of a packet of data and send it along one of two paths. The first path went upstairs to an equipment room belonging to Cogent, a wholesale Internet provider that serviced cities from San Francisco to Kiev. A yellow wire passed through a utility shaft, came through a wall, and plugged into Cogent's equipment, itself connected to electronic colleagues in Chicago and Minneapolis. This building was Cogent's only "point of presence" in all of Wisconsin, the only place Cogent's express train stopped; that's why Auer's company was here, and all the others. The second cable went to Time Warner, whose wholesale Internet division provided an additional connection—a backup, plugging Auer's piece of the Internet into all the rest.

Taken as a whole, the building seemed a labyrinth, packed with a hundred years of twisted cables and broken dreams. Yet in its particularity, this part of the Internet—Auer's part—was strikingly legible; it wasn't an endless city at all but a simple fork in the road. I asked Auer what happened after here, and he shrugged. "I care about where we can talk to Cogent or Time Warner, which means this building. Once it's here it's really out of my hands." For about twenty-five thousand Wisconsinites, this was the source. Their Internet went this way and it went that way: two yellow cables leading, eventually, to the world. Every journey—physical and virtual—begins with a single step.

———————

A few weeks later I went to Washington to visit TeleGeography's offices, for a better sense of how Krisetya drew a clear map of the Internet's mushy layer cake. But the night before I left, New York was hit by a blizzard, and I emailed Krisetya to let him know I'd be arriving later than expected. As the train moved south across New Jersey the snow began to dwindle, so that by the time we pulled into Washington the blanket of white I had left in New York had given way to clear gray sky and dry sidewalks. It was as if over the course of the ride the veil that had descended upon the landscape had just as quickly been lifted. Arriving in DC, I opened my laptop in the center of Union Station's great neoclassical hall to log into a café's wireless network and send off an email to California. A few minutes later, standing on the Metro platform, I thumbed a message to my wife saying that, despite New York being shut down by the snow, I had made it to Washington (and we'll see about getting back).

I share all these quotidian details of travel because on that day my senses were unusually attuned to the networks that surrounded me, both visible and invisible. Maybe it was the way the snow had drawn a new outline around the world's familiar shapes, while slowing my progress past them. Or maybe it was just the early morning hour and the fact that I had maps on the brain. But as the train was sliding across the elbow of New Jersey, ducking out of the storm, I could imagine the emails following (albeit faster) along the same path. I had recently learned that many of the fiber-optic routes between New York and Washington were lain along the railroad tracks, and I could begin to imagine the route my email to California had taken: it might have shot back the way I'd come, to New York, before heading cross-country, or it could have continued farther west to Ashburn, Virginia, where there was an especially significant network crossroads. The exact route of that email didn't matter; what did was that the Internet no longer seemed infinite. The invisible world was revealing itself.

In a neighborhood of staid lobbyists and wood-paneled law firms, TeleGeography's K Street office stands out for its lime-green walls, exposed ceilings, and translucent cubicle dividers. The front door pivoted creatively on its center point. Maps lined the walls, of course. On one, Spain had been adorned with a Groucho Marx mustache, a remnant of a recent holiday party. Krisetya welcomed me into his office, the desk piled high with books about information design. When he joined TeleGeography in 1999, he was put right to work on the company's first big report, *Hubs + Spokes: A TeleGeography Internet Reader.* It was groundbreaking. Before, there were geographic maps showing the networks operated by individual corporations or government agencies, and there were "logical" diagrams of the whole Internet, like a subway map. Neither gave a strong sense of how the Internet adhered and diverged from the real-world geography of cities and countries. What places were *more* connected? Where were the hubs?

Krisetya began looking at new ways of portraying that combination of the geopolitical and networked worlds. He blended the outlines of the continents with diagrams of the networks, "always layering something abstract on top of something that's familiar, always looking to give it more meaning." Other kinds of maps had long struggled with the same issues—like airline routes or subways. In both cases, the end points were more important than the paths themselves. They always had to balance the workings of the system internally with the external world it connected. London's Tube map might be the height of the genre: a geographical fiction that pushes and pulls at real-world creations, leaving in its wake a kind of alternate city that's become as real as the true one.

On his maps, Krisetya portrayed this by showing the most heavily trafficked routes between cities, such as between New York and London, with the thickest lines—not because there were necessarily more cables there (or some single, superthick cable) but because that was the route across which the most data flowed. This was an insight that dated back to that first report. "If you look inside the Internet cloud a fairly distinct hub-and-spoke structure begins to emerge at both an operational (networking) and physical (geopolitical) level," it explained. The Internet's structure "is based upon a core of meshed connectivity between world cities on coastal shores—Silicon Valley, New York and Washington, DC; London, Paris, Amsterdam and Frankfurt; Tokyo and Seoul." And it still is.

Today's version—the one TeleGeography calls the *GIG*—is the bible for big telecommunications companies. The key to its approach is still to look at Internet traffic as concentrated between powerful cities. TeleGeography breaks down the nebulous cloud into a clear system of point-to-point communications, of segments. Contrary to its ostensible fluidity, the geography of the Internet reflects the geography of the earth; it adheres to the borders of nations and the edges of continents. "That's the nugget of our approach," Krisetya explained to me in his office, sounding like a college tutor. "We always put much more emphasis on the actual geography than the connections in between. In the beginning, that's what we were more familiar with. When the Internet was still very much abstract, we knew where the two end points were, even if we didn't understand how this was all being built."

That had a certain clarity. The world is real; London is London, New York is New York, and the two usually had a lot to say to each other. But I was still hung up on what seemed a simple question: What, physically speaking, *were* all those lines? And where precisely did they run? If TeleGeography properly understood the Internet as being "point to point," what and where were the points?

For their part, TeleGeography's analysts don't go out into the world with a GPS and a sketchpad. They don't attach sensors to the Internet to measure the speed of the bits passing by, like a water meter. Their process is quite low tech: they distribute a simple questionnaire to telecom executives, requesting information about their networks in exchange for the promise to keep it confidential and to share the aggregated information with them. And then TeleGeography asks the Internet itself.

To see how, Krisetya dropped me off at the tidy desk of Bonnie

Crouch, the young analyst responsible for gathering and interpreting TeleGeography's data on Asia. The diplomatic work of wrangling and cajoling the information from the telecom carriers was finished, and the responses loaded into TeleGeography's database. Crouch's job was to confirm what the carriers said, based on the Internet's actual traffic patterns. Cartographers talk about "ground truth": the in-person measurements used to check the accuracy of the "remote sensing"—which in contemporary mapmaking usually means aerial or satellite photographs. TeleGeography had its own way of checking the "ground truth" of the Internet.

When I enter an address into my browser, a thousand tiny processes are set in motion. But in the most fundamental terms, I'm asking a computer far away to send information to a computer close by, the one in front of me. Browsing the web, that typically means a short command—"send me that blog post!"—is volleyed back with a far larger trove, the blog post itself. Behind the URL—say, www.mapgeeks.com—is a self-addressed envelope with the instructions that connect any two computers. Every piece, or "packet," of data traveling across the Internet is labeled with its destination, known as an "IP" address. Those addresses are grouped into the equivalent of postal codes, called "prefixes," given out by an international governing body, the Internet Assigned Numbers Authority. But the routes themselves aren't assigned by anyone at all. Instead, each router announces the existence of all the computers and all the other routers "behind" it, as if posting a sign saying THIS SECTION OF THE INTERNET OVER HERE. Those announcements are then passed around from router to router, like a good piece of gossip. For example, Jon Auer's router in Milwaukee is the doorway to his twenty-five thousand customers, grouped into just four prefixes. It announces its presence to the two neighboring routers, belonging to Cogent and Time Warner. Those two neighboring routers make a note of it, and then pass the word on to their neighbors—and so on, until every router on the Internet knows who's behind whom. The complete aggregate list of destinations is known as the "routing table." At the end of 2010 it had nearly four hundred thousand entries and was growing steadily. The whole thing is typically stored in the router's internal memory, while a compact flash card, like the kind used by digital cameras, keeps the operating code. Auer buys his on sale at the local drugstore.

Two things surprised me about this. The first is that every IP address is by definition public knowledge; to be on the Internet is to want to be found. The second is that the announcement of each route is based wholly on trust. The Internet Assigned Numbers Authority gives out the prefixes, but anyone can put up a sign pointing the way. And sometimes that does go horribly wrong. In one well-known incident in February 2008, the Pakistani government instructed all Pakistani Internet providers to block YouTube, because of a video it deemed offensive. But an engineer at Pakistan Telecom, receiving the memo at his desk, misconfigured his router, and rather than removing the announced path to YouTube, he announced it himself—in effect declaring that he *was* YouTube. Within two and a half minutes, the "hijacked" route was passed to routers across the Internet, leading anyone looking for YouTube to knock on Pakistan Telecom's door. Needless to say, YouTube wasn't in there. For most of the world, YouTube wasn't available at all for nearly two hours, at which point the mess was sorted out.

It sounds preposterously loose and informal. But it strikes at the core of the Internet's fundamental openness. There's a certain amount of vulnerability involved with being a network on the Internet. When two networks connect, they have to trust each other—which also means trusting everyone the other one trusts. Internet networks are promiscuous, but their promiscuity is out in the open. It's free love. Jon Postel, the longtime administrator of the Internet Assigned Numbers Authority, put this into a koan, a golden rule for network engineers: "Be conservative in what you send, be liberal in what you accept."

For TeleGeography this means everything is out in the open, for those who know how to see it. The company uses a program called Traceroute, originally written in 1988 by a computer scientist at the Lawrence Berkeley National Laboratory. He had gotten fed up, as he put it in a mailing list message to his colleagues, trying to figure out "where the !?*! are the packets going?" and worked up a simple program that traced their paths. Enter in an IP address and Traceroute will feed back a list of the routers traversed to reach it, and the time (in milliseconds) elapsed in the journey between each one. TeleGeography then takes it one step further. It carefully selects fifteen locations around the world, looking especially for "dead-end" places with only a few paths out to the rest of the Internet—Denmark's Faroe Islands, for example. It then searches for websites there hosting a copy of the Traceroute program (often a university computer science department), and directs those fifteen Traceroute hosts to query more than twenty-five hundred "destinations," websites carefully chosen because they could reasonably be expected to actually live on a hard drive in the place where they say they live. Jagiellonian

University in Poland, for example, is unlikely to host its website in, say, Nebraska. That meant TeleGeography in Washington was asking a computer science department in Denmark to show how it was connected to a university in Poland. It was like a spotlight in Scandinavia shining on twenty-five hundred different places around the world, and reporting back on the unique reflections. TeleGeography's trick was finding real-world corners and dead ends, thereby minimizing the number of possible paths.

Added all together, the fifteen hosts TeleGeography selected query twenty-five hundred destinations yielding more than twenty thousand journeys across the Internet—and, inherently, around the earth. Quite a few of these journeys are never completed; the traces conk out, lost in the ether. The whole set takes several days, not because TeleGeography has a slow computer, or even a slow Internet connection. Rather, those days represent the aggregate duration of all those thousands of trips, milliseconds piled upon milliseconds in which the explorer packets are crisscrossing the earth. And I don't mean "crisscrossing" idly. These paths are by no means random or imaginary. Each packet—a clump of math, in the form of electrical signals or pulses of light—moves along very specific physical pathways. The whole point of each traceroute is to identify that specificity, that singular record of a journey. Theoretically you could divide up the task of querying each traceroute among multiple computers, but there's no way to rush the traces themselves, no more than you can rush the speed of light. The time the packets take on their journey is the time they take. Each recorded journey is like a series of tiny postcards from around the globe. TeleGeography then layers the tens of thousands of them as if they were strands of papier-mâché, until the patterns emerge.

Crouch and the other analysts then parse the routes by hand. "Any particular country of interest?" she asked me, with the geographic expansiveness I was quickly learning to love about Internet people. I told her to pick whichever she knew best, and she chose Japan—dodging the ambiguity of China's networks. On her screen, a long list of jumbled letters and numbers, like a phone book without the names, scrolled down. Each grouping represented the results of a single trace—from the Faroes to Hokkaido, for example. Each individual line represented a single router: a lonely machine in a cold room, studiously forwarding packets. Over time the codes had become familiar to Crouch, like London's streets are to a cabdriver. "You begin to get a feel for how companies name their routers," she said. "Like that one's going from SYD to HKG—the airport codes for Sydney to Hong Kong. And the carrier did tell us it's running that route, so we don't need to worry about it." Her goal in reading these lists was to confirm that the carriers are operating the routes they say they are, and, with a more subjective eye, to make a judgment about the amount of traffic on that route. "Our research gives us all the pieces of the puzzle: the bandwidth, the Internet capacity, some of the pricing information. The gaps in between we can fill in with some reasonable accuracy."

It occurred to me that Crouch was part of the small global fraternity that knows the geography of the Internet the way most people know their hometowns. Her boss, a Texan named Alan Mauldin who improbably led TeleGeography's analyst team from his home in Bratislava, possessed one of the best mental maps of the physical infrastructure of the Internet. I'd spoken with him before coming. "I don't need to look at a map," he told me over Skype. "I have in my mind, and I can almost note, which cables connect everywhere in the world." Rather than maps of the Internet, his study in Slovakia was decorated with antique maps of Texas. "I suppose it is kind of like the *Matrix,* where you can see the code. I don't even have to think about it anymore. I can just see where it's going. I know what city the router is in, and where the packet's going. It's the weirdest thing. But it's easy to just fly through it all, once you know what to look for."

Yet what's so striking to me—and so often overlooked—is that each router is inherently *present*. Each router is a singular waypoint, a physical box, in a real place, on a packet's journey across this real earth. Two billion people use the Internet from every country on earth; airplanes have Wi-Fi; astronauts browse the web from space. The question "Where is the Internet?" should seem meaningless, because *where isn't it?* And yet, standing over Crouch's shoulder, watching her identify the coded name of individual machines in a city on the other side of the world, the Internet didn't seem infinite at all. It seemed like a necklace strung around the earth. Forming what pattern? Did it look like the route maps in the back of an airline magazine? Or was it more chaotic, like a bowl of spaghetti or the London Underground? Before, I'd imagined the Internet as something organic, beyond human design, like an ant colony or a mountain range. But now its designers seemed present, not an innumerable crowd but a tidy contact list on a laptop in Washington. So who were they then? Why did they lay their networks there? Where did it all begin?

2

A Network of Networks

I wanted to know where the Internet started, but the question turned out to be more complicated than I'd imagined. For an invention that dominates our daily lives—acknowledged as an epoch-making transformative force across global society—the Internet's history is surprisingly underwritten.

The serious book-length histories all seem to have been published in 1999, as if the Internet were finished then—as if the Internet were finished now. But more than that timing, they each seem to have their own heroes, milestones, and beginnings. The Internet's history, like the network, was itself distributed. As one historian of historians put it (writing in 1998), "the Internet lacks a central founding figure—a Thomas Edison or a Samuel F. B. Morse." I should have known things wouldn't be so clear-cut when the author of *Inventing the Internet*, widely considered the most authoritative among them, began by suggesting that "the history of the Internet holds a number of surprises and confounds some common assumptions." I felt like a guy wandering around a party he wasn't invited to, asking who the host was, and nobody knew. Or maybe there wasn't a host at all? Maybe the problem was more philosophical than that? The Internet had a chicken-and-egg thing going on: If the Internet is a network of networks, then it takes two networks to make an Internet, so how could one have been the first?

Needless to say, all this did not inspire confidence. I had set out in search of the real, the concrete, the verifiable, but I was greeted at the door by the historiographic equivalent of a comments thread. My question had to be narrower, more rooted in time and place. It was about the object. "Not ideas about the thing but the thing itself," as Wallace Stevens wrote. Not, where did the Internet begin? But, where was its first box? And that, at least, was clear.

In the summer of 1969, a machine called an interface message processor, or IMP, was installed at the University of California–Los Angeles, under the supervision of a young professor named Leonard Kleinrock. He's still there, a little less young, but with a boyish smile and a website that seemed to encourage visitors. "You'll want to meet me in my office," he replied when I emailed. "The original site of the IMP is just down the hall." We made arrangements. But it wasn't until I settled into my cramped seat on the plane to Los Angeles, surrounded by tired consultants in wrinkled shirts and aspiring starlets in sunglasses, that the full implications of my journey sank in: I was going to visit the Internet, flying three thousand miles on a pilgrimage to a half-imagined place. And what did I expect to find? What, truly, was I looking for?

I suppose all pilgrims feel that way at some point. We are optimistic creatures. In Judaism, the Temple Mount in Jerusalem is the place from which the whole world expanded, the place closest to God, and the most important place of prayer. For Muslims, the small cube building in Mecca known as the Kaaba is the holiest place, so dominant in the psychic geography of the devout that they face it to pray five times a day, wherever they are in the world, even flying across the ocean on an airplane. Every cult, group, team, gang, society, guild—whatever—has its significant place, marked with memory and meaning. And most of us also have our own individual places: a hometown, stadium, church, beach, or mountain that looms epically above our lives.

Yet this significance is always in a way personal, even if millions collectively share it. Philosophers like to point out that "place" is as much within us as without us. You can demarcate a place on a map, pinpoint its latitude and longitude with global positioning satellites, and kick the very real dirt of its very real ground. But that's inevitably going to be only half its story. The other half of the story comes from us, from the stories we tell about a place and our experience of it. As the philosopher Edward Casey writes, "Stripping away cultural or linguistic accretions, we shall never find a pure place lying underneath." All we shall find instead are "continuous and changing qualifications of particular places." When we travel, we fix a place's meaning in our minds. It is in the eyes of a pilgrim that a holy site becomes holiest. And in being there, he affirms not only the place's significance but also his own. Our physical place helps us better know our psychic place—our identity. But did that hold true for me, on the path to the Internet? I longed to see its most significant places, but were its places really places at all? And if they were, was the Internet close enough to religion—a way of understanding the world—that seeing those places would be meaningful?

The question got more complicated the next morning in Los Angeles. I woke up at dawn, my body clock still set to New York, in an enormous hotel near the airport, with a mirrored glass façade and a view of the runway. I stood in front of the window watching a line of jets land on top of their shadows. Inside, nearly every surface had a little folded cardboard sign that indicated the branded touches that made the room conform to the hotel chain's international standards: the "Suite Dreams®" bed, the "Serenity Bath Collection™," the "Signature Service." Nothing was singular or local; everything came from far away, at the direction of a global corporation. The novelist Walter Kirn calls this "Airworld"—these nonplaces of airports and their surroundings. I tried to get a little postmodern kick out of it all and summon my inner Ryan Bingham, the protagonist of Kirn's novel *Up in the Air* (played by George Clooney in the movie) who only feels at home here in this homogeneous, if admittedly comfortable, world—even if "cities don't stick in my head the way they used to." But I was hollow. On my way to the Internet I was already climbing a steep slope toward the singular and the local. It was frustrating to find the ostensibly real places blurring into each other as well. I had come to Los Angeles to try to bring the network back into the world—but instead, the world seemed to have succumbed to the logic of the network.

But I shouldn't have worried. At UCLA that afternoon, the moment of the Internet's physical birth came vividly into focus, rooted in a very specific place. On the quiet Saturday afternoon of the Labor Day weekend in 1969, a small crowd of computer science graduate students had gathered in the courtyard of Boelter Hall with a bottle of champagne. Standing in the same spot, I conjured the scene. The occasion was the arrival of their grand and expensive new gadget, coming that day from Boston by air freight: a modified and military-hardened version of a Honeywell DDP-516 minicomputer—"mini," at the time, meaning a machine that weighed nine hundred pounds and cost $80,000, the equivalent of nearly $500,000 today. It was traveling from the Cambridge, Massachusetts, engineering firm of Bolt, Beranek and Newman, possessor of a $1 million Department of Defense contract to build an experimental computer network, known as the ARPANET. Among Bolt's many customizations was a new name for the machine: the Interface Message Processor. The particular one then wending its way up the hill to the UCLA campus was the very first: IMP #1.

The graduate students were mostly my parents' age, born in the last days of World War II—proto–baby boomers, in their midtwenties at the time—and I can blur in some family photos of the era. It was the summer of Woodstock and men on the moon, and even computer scientists wore their hair shaggy and their pants legs wide. One probably had a button with the word RESIST on it next to a question mark, the scientific notation for electrical resistance and a popular antiwar symbol for engineers. They all knew that their funding, $200,000 to UCLA alone supporting forty grad students and staff, came from the Department of Defense. But they also all knew that what they were building wasn't a weapon.

The ARPANET project was managed by the Department of Defense's Advanced Research Projects Agency (ARPA), founded in the wake of Sputnik's launch to support scientific research, esoteric stuff far out on the technological frontier. ARPANET certainly qualified. There had been few attempts to connect computers at continental-scaled distances, let alone create an interconnected network of them. If somewhere deep in the Pentagon a four-star general had the grim notion that the nascent ARPANET might evolve into a communications network that could survive a nuclear war—a popular myth about the origins of the Internet—this group was insulated from it. And anyway, they ignored it. They were consumed by the technical challenges arriving inside that moving van, by new wives and babies, and by the infinite possibilities of computer communications. Consumed, that is, by peaceful intentions.

Boelter Hall was new and shiny then, like much of Los Angeles itself. Built in the early 1960s to house the rapidly expanding engineering department, its stripped modernist lines were the height of architectural fashion, suitable to the cutting-edge work going on inside—not unlike the new biomolecular science building that now towers over it next door. These days Boelter is a bit rough around the edges, with worn-out sunshades over the windows and rusting steel balcony railings facing a courtyard of full-grown eucalyptus trees. The IMP's welcoming party would have stood here beneath them in the shade, on a hot Southern California day. Long before cell phones, they would have guessed at the timing of the truck's progress from the airport. A forklift waited nearby, ready to lift the massive machine up into the building. Did they sip their champagne from Styrofoam cups? Snap pictures with one of the inexpensive new Japanese cameras that had recently begun to be imported? (If so, they're long lost.) The excitement of the occasion would have been unmistakable, even if the full historic implications were not: this was the first piece of the Internet.

But while the grad students were celebrating outside, their professor was stuck upstairs, alone in the large office he had recently expanded in a fit of empire building, shuffling papers on a Saturday afternoon. This I can picture precisely, because when I walked in forty-one years later, Leonard Kleinrock was still sitting there, sprightly at seventy-five, wearing a starched pink shirt, black slacks, and a BlackBerry clipped to a polished leather belt. His face was tanned and his hair was full. A brand-new laptop was open on his desk and he was yelling into a speakerphone: "It's not catching!"

On the other end, the disembodied voice of a tech support person responded slowly and patiently. *Click here. Now click there. Type this in.* Kleinrock looked over the top of his reading glasses and waved me toward a chair. Then he clicked. And clicked again.

Now try, the voice said.

He winced. "It says I'm not connected to the Internet. That's what it says!" Then he laughed so hard his shoulders shook.

Kleinrock is the father of the Internet—or rather, *a* father, as success has many. In 1961, while a graduate student at MIT, he published the first paper on "packet switching," the idea that data could be transmitted efficiently in small chunks rather than a continuous stream—one of the key notions behind the Internet. The idea was already in the air. A professor at the British National Physical Laboratory named Donald Davies had, unbeknownst to Kleinrock, been independently refining similar concepts, as had Paul Baran, a researcher at the RAND Corporation in Los Angeles. Baran's work, begun in 1960 at the request of the US Air Force, was explicitly aimed at designing a network that could survive a nuclear attack. Davies, working in an academic setting, merely wanted to improve England's communications system. By the mid-1960s—by which time Kleinrock was at UCLA, on his way toward tenure—their ideas were circulating among the small global community of computer scientists, hashed out at conferences and on office chalkboards. But they were only ideas. No one had yet fit the pieces of the puzzle together into a working network. The fundamental challenge these network pioneers faced—and the one that remains at the heart of the Internet's DNA—was designing not just a network but a network of networks. They weren't only trying to get two or three or even a thousand computers talking, but two or three or a thousand different kinds of computers, grouped in all sort of ways, spread far and wide. This metalevel challenge was known as "Internetworking."

It took the Department of Defense to bring teeth to it. In 1967, a young computer scientist named Larry Roberts—Kleinrock's MIT office mate—was recruited to ARPA specifically to develop an experimental nationwide computer network. The next July, he sent out a detailed request for proposals to 140 different technology companies to build what he at first called the "ARPA net." It would begin at four universities, all in the west: UCLA, Stanford Research Institute, the University of Utah, and the University of California–Santa Barbara. The geographic bias wasn't an accident. Connecting university computers together was a threatening idea—each school would inherently have to share its prized, and already overused, machine. The East Coast schools tended to be more conservative, or at least less susceptible to Roberts's ability to influence them with his control over their ARPA funding. California already had a burgeoning technology culture and big universities, but the nascent ARPANET's West Coast beginnings had as much to do with a cultural appetite for new ideas.

Under Kleinrock's direction, UCLA would have the added task of hosting the Network Measurement Center, responsible for studying the performance of this new creation. The reason for that was as much personal as professional: not only was Kleinrock the reigning expert in network theory but Roberts trusted his old friend. If Bolt, Beranek and Newman's job was to build the network, Kleinrock's would be to break it, to test the limits of its performance. It also meant that UCLA would receive that first IMP, to be installed between the computer science department's big, shared computer, called a Sigma-7, and the specially modified phone lines to the other universities that AT&T had readied for when there eventually was a network. For its first month in California, IMP #1 stood alone in the world, an island awaiting its first link.

"Do you want to see it?" Kleinrock asked, jumping excitedly from his chair. He led me across the hall to a small conference room, not fifteen feet from his desk. "This is it—a beautiful machine! A really magnificent machine!" The IMP looked—like all famous things—exactly like it does in photographs: refrigerator-sized, beige, steel, with buttons on the front, like a file cabinet dressed up as R2-D2. He opened and closed the cabinet and twisted some dials. "Military hardened, built out of a Honeywell DDP-516, state of the art at the time." I said I had begun to notice that the Internet had a smell, an odd but distinctive mix of industrial-strength air conditioners and the ozone released by capacitors, and we both leaned in for a whiff. The IMP smelled like my grandfather's basement. "That's mildew," Kleinrock said. "We should close the door and it'll cook it up."

The IMP's current situation certainly lacked in ceremony, shoved as it was into the corner of a small conference room, with mismatched chairs and faded posters on the walls. A stack of paper coffee cups poked out of a plastic bag. "Why is it sitting here?" Kleinrock said. "Why is it not in a wonderful showcase somewhere on campus? The reason is nobody took this machine as being important. They were going to throw it out. I had to rescue it. Nobody recognized its value. I said, 'We gotta keep this thing, it's important!' But you're never a philosopher in your own country."

But that was changing. A history graduate student at UCLA had recently clued into the historical significance of Boelter Hall and the IMP and had begun assembling archival materials. After years of pleading with the university administration, Kleinrock had finally gathered support for the construction of the Kleinrock Internet Heritage Site and Archive. It would commemorate not only the IMP itself but the historical moment. "It was amazing, this group of really smart people collected in the same time and the same place," Kleinrock said. "It happens, it's sort of periodic, when you get this kind of golden era." Indeed, the group assembled in his lab that fall formed a core group of Internet hall of famers, notably Vint Cerf (now "Chief Internet Evangelist" at Google), who cowrote the Internet's most important operational code—what is known as the TCP/IP protocol—with Steve Crocker, also Kleinrock's student, and Jon Postel, who managed the Internet Assigned Numbers Authority for years and was a key mentor to an entire generation of network engineers.

The museum would be built in room 3420, where the IMP had been installed from Labor Day 1969 until it was decommissioned in 1982. We walked down the hall to see it. "The IMP was against this wall here," Kleinrock said, slapping the white paint, "but the room has been reconfigured. The ceiling is new, the floor is new—we had a raised floor for air-conditioning." We peered behind a steel storage cabinet to see if the original phone jack might still be there—the first few feet of the Internet's first route—but it wasn't. There was no plaque, no historic display, and there were certainly no tourists, not yet. Kleinrock's hope was to restore the room to the way it looked in 1969, which I imagined would become something like Graceland, frozen in time, with the IMP and an old rotary telephone and photographs of men in heavy-framed glasses and with slicked-down hair. "To put a wall up here and put a doorway there is $40,000, and we have a $50,000 budget for this and the archivist," Kleinrock said. "So I'm going to have to donate a lot of money, I think. It's okay. It's a good cause."

As we talked, an undergraduate computer lab was in progress in the room, with students touching soldering irons to green circuit boards, their cell phones set on the desks in front of them, while a teaching assistant barked instructions. No one even glanced at us. Kleinrock was one of the Internet's earliest masterminds, but to the nineteen-year-olds in here, whose lives were fully shaped by its presence—Internet Explorer came out before they would have learned to read—he faded into the woodwork. Almost literally. This wasn't a holy site, it was a classroom—far less of a tourist attraction than Ryan Seacrest's house, not far away. So what was I doing here?

Boelter Hall was where the Internet had once been fully containable, in stark contrast with its current sprawl. And Kleinrock was still there, embodying that history in his memory. But I could have reached him on the phone, we could have video-chatted. But I had cast my net in the waters of experience, having chosen (for example) to ignore the photograph of room 3420 that shows up in a Google image search, and instead come see it for myself. That afternoon, when I had arrived early for my appointment with Kleinrock, I sat on a ledge outside Boelter Hall eating a bag of chips, fiddling with my cell phone. My wife had just emailed a video of our baby taking her first crawling steps, a video that loaded vividly on the small screen and pulled me back, in my head, to New York. I had come to visit the first node of the Internet, but one of its most recent nodes—the one I carried in my pocket—had distracted me. If the Internet was a fluid new world distinct from the old physical world, Boelter Hall struck me as a place where the two met, forming an unusually visible seam. Except the essence I sought was diluted by the evolution of the thing that it created. Here was the shiny new device, connected everywhere; there was the ancient machine in the wood case, smelling of mildew. What, really, was the difference? The IMP here was the real thing: not a replica or a model—or a digital image. That's why I was here: to hear the details from Kleinrock himself and note the color of the walls, but also to thumb my nose at the immediate reproducibility of everything else. The place itself couldn't be blogged and reblogged—and I confess that I was a little drunk on the irony of that. In his 1936 essay, "The Work of Art in the Age of Mechanical Reproduction," Walter Benjamin describes the fading importance of an object's "aura," its unique essence; and here I was in search of the aura of the thing that threatened to destroy the idea of "aura" once and for all.

I asked Kleinrock about this: Why isn't *essence* a word we typically talk about in the context of the Internet itself? It's more often the opposite that thrills us: the network's ease at instant reproduction, its ability to make things "viral," with the consequence of threatening not only the aura but also our desire for it—leading us to watch a concert through a smartphone screen. "For the same reason people don't know when it was created or where it started, or what the first message was," he said. "It's an interesting psychological and sociological commentary that people are not curious about it. It's like oxygen. People don't ask where oxygen comes from. I think the students today miss a lot because they don't take things apart. You can't take this apart"—he rapped his new laptop. "Where's the physical experience? Unfortunately it's gone. They have no idea how this thing works. When I was a kid building radios I knew what I was dealing with; I knew how things worked and why they worked that way." The lab then in progress in room 3420 was an exception, the one time the computer science students got their hands dirty.

I asked Kleinrock about some of the mementos lying around his office. From a small gray archival box resting on top of a file cabinet he pulled out the original log recording the moment when UCLA's IMP first connected with IMP #2, installed at Stanford Research Institute, late on the Wednesday evening of October 29, 1969. The notebook was tan, with "IMP LOG" written in sloppy marker on its cover. You can see it on Kleinrock's website, of course. "That's the most precious document on the Internet," he said. "There is someone putting together these archives now, and they shoot me every time I touch this. They're the ones who gave me this box." He opened it up and began to read the entries:

SRI called, tried debug test, but it didn't work.

Dan pushed some buttons.

"The important one is here—October twenty-ninth. I shouldn't be touching these pages, but I can't resist! There it is." In blue ballpoint pen, the words running over two lines and beside the time code 22:30, was written:

Talked to SRI host to host.

It is the sole documentary evidence of the ARPANET's first successful transmission between sites—the moment of the Internet's first breath. I nervously kept my hands in my lap. "If anybody comes to steal it, here it is!" Kleinrock said. "There's also a copy of my dissertation."

Then he turned nostalgic. "In those early days, none of us had any idea what it would become. I had a vision, and I got a lot of it right. But what I missed was the social side—that my ninety-nine-year-old mother would be on the Internet when she was alive. That part eluded me. I thought it was going to be computers talking to computers or people talking to computers. That's not what it's about. It's about you and me talking."

I reminded him that we had closed up the IMP to let the smell "cook up," and we again crossed the hall to pay our respects. Kleinrock opened the cabinet. "Here," he said. "Yeah. Mmmm. Get your nose down there." I leaned in, as if toward a flower. "Smell that? There are components here, there's rubber. This is the stuff when I was a kid I used to cannibalize old radios, with vacuum tubes, and I would smell the solder a lot, the resin." I recalled the electronics class I took as a third grader, after school. We made LEDs that blinked in a pattern. I spend my days connected to electronic machines but I'd hardly smelled it since.

"You can't record that," Kleinrock said. "Yet. One day you will."

The Internet's adolescence was protracted. From the ARPANET's birth at UCLA in 1969 until the mid-1990s, the network of networks crept slowly outward from universities and military bases to computer companies, law firms, and banks, long before it found its way to the rest of us. But in those long early years there really wasn't much of it to speak of. For a quarter century, Kleinrock and his colleagues were like explorers, staking the flag of the nascent Internet on a series of far-flung colonies, connected only tenuously with one another, and often not at all with their immediate surroundings. The Internet was thin on the ground.

The early maps of the ARPANET frequently published by Bolt, Beranek and Newman show just how thin. They look like constellation charts. In each version, an outline drawing of the

United States is overlaid with black circles indicating each IMP, linked by razor-straight lines. The ARPANET began life as the Little Dipper, scooping up a piece of California, its handle in Utah. By the summer of 1970, it had expanded east across the country to add MIT, Harvard, and Bolt's Cambridge offices. Washington didn't appear until the following fall. By September 1973, the ARPANET went international, with the establishment of a satellite link to University College London. By the end of the decade, the network's geography was fully entrenched around four regions: Silicon Valley, Los Angeles, Boston, and Washington. New York City hardly appeared at all, with only NYU winning a colony. Only a few scattered nodes dotted the middle of the United States. True to its philosophical roots as a doomsday-era communications system, the ARPANET was strikingly deurbanized and decentralized. It had no special places, no monuments. Physically speaking, there were IMPs like the one down the hall from Kleinrock's office, linked by always-on phone connections provided under special terms by AT&T. It existed in spare classrooms of university computer science departments, within outbuildings on military bases, and across the copper lines and microwave links of the existing telephone network. The ARPANET wasn't even a cloud. It was a series of isolated outposts strung together by narrow roads, like a latter-day Pony Express.

No doubt there was serious research going on, but the ARPA-NET's use as a communications tool retained an air of novelty. In September 1973, a conference at Sussex University, in Brighton, England, brought together computer scientists from around the world who were each developing their own government-sponsored computer networks. Since the ARPANET was the largest, a special demonstration link was established back to the United States. It hadn't been an easy thing to arrange. A telephone line had to be activated between one of the ARPANET nodes in Virginia and a nearby satellite antenna. From there, the signal was bounced off an orbiting satellite to another earth station in Goonhilly Downs in Cornwall, then onward through a telephone link to London, and finally into Brighton. It was less a technological marvel than what engineers like to call a "kludge," a temporary and tenuous link across the ocean.

But history remembers it for more prosaic reasons. In a story that has become legend, when Kleinrock arrived home in Los Angeles from the conference, he realized he'd forgotten his electric razor in the Sussex dormitory bathroom. Logging into the ARPANET from his UCLA computer terminal, he entered the command WHERE ROBERTS, which told him if his friend Larry Roberts—a well-known workaholic and insomniac—was logged in as well. Sure enough he was, wide awake at 3:00 A.M. Using a rudimentary chat program—"clickety clickety clack," as Kleinrock describes it—the two friends made arrangements to send home the razor. That kind of communication was "a bit like being a stowaway on an aircraft carrier," as the historians Katie Hafner and Matthew Lyon describe it.

The 1970s ARPANET was US government property, linking defense researchers either within the military itself or at ARPA-funded university departments. But socially the ARPANET was a small town. The 1980 edition of the ARPANET directory is a canary-colored perfect-bound book, about the thickness of a fall fashion magazine. It lists the five-thousand-odd names of everyone on the ARPANET, with their postal addresses, the lettered code for their nodes, and their email addresses—absent the ".com" or ".edu," which wouldn't be invented for another few years. Kleinrock is in there, of course, with the same office address and phone number as today (although his area code, zip code, and email address have all changed). Sharing the page with him are computer scientists at MIT, University College London, and the University of Pennsylvania; a commander at the Army Communications Research and Development Command at Fort Monmouth, New Jersey; and the chief of the Strategic Studies Programming Division at Offutt Air Force Base in Nebraska—famous as the manufacturing site of the *Enola Gay,* the primary Cold War–era nuclear command center, and the place at which President Bush temporarily sought refuge on 9/11.

The ARPANET was like that: an accidental meeting place for academics and high-tech soldiers, brought together under the umbrella of computer networking. Inside the front flap of the directory is a logical map of the ARPANET, with each node labeled in tiny print and connected together with thick and thin straight lines, like an elaborate and convoluted flowchart. Every computer on the ARPANET fits easily on the page. But that intimacy wouldn't last.

By the early 1980s, the big computer companies—like IBM, XEROX, or Digital Equipment Corporation—and large government agencies—like NASA and the Department of Energy—were running their own independent computer networks, each with its own acronym. High-energy physicists had HEPnet. Space physicists had SPAN. Magnetic fusion researchers had MFENET. A handful of European networks had also emerged, including EUnet and EARN (the European Academic Research Network). And there were a growing number of regional academic networks, named like the twelve sons of Mr. and Mrs. Net: BARRNet, MIDnet, Westnet, NorthWestNet, SESQUINet.

The trouble was, all those networks weren't connected. While stretching nationwide and occasionally across the ocean, they operated in effect as private highways overlaid on the public telephone system. They overlapped geographically, sometimes serving the same university campuses. And they even might have overlapped physically, sharing the very same long-distance telephone cables. But in networking terms they were "logically" distinct. They were disconnected—as separate as the sun and the moon.

That remained the case until New Year's Day 1983 when, in a transition years in the planning, all the host computers on the ARPANET adopted the electronic rules that remain the basic building block of the Internet. In technical terms, they switched their communications protocol, or language, from NCP, or "Network Control Protocol," to TCP/IP, or "Transmission Control Protocol/Internet Protocol." This was the moment in the Internet's history when the child became father to the man. The changeover, led by the engineers at Bolt, Beranek and Newman, kept dozens of system administrators tied to their desks on New Year's Eve, struggling to make the deadline—leading one to commemorate the ordeal by making I SURVIVED THE TCP/IP TRANSITION buttons. Any node that did not comply was cut off until it did. But once the dust had settled several months later, the result was the computing equivalent of a single international language. TCP/IP went from a dominant dialect to an official lingua franca.

As the historian Janet Abbate notes, the changeover marked not just an administrative shift but a crucial conceptual one: "It was no longer enough to think about how a set of *computers* could be connected; network builders now also had to consider how different *networks* could interact." The ARPANET was no longer a walled garden with an official government directory of participants, but rather had become just one network among many, linked together into an "Internetwork."

The New Year's 1983 standardization of TCP/IP permanently fixed the Internet's distributed structure, ensuring to this day its lack of central control. Each network acts independently, or "autonomously," because TCP/IP gives it the vocabulary to *inter*act. As the author and Columbia law school professor Tim Wu points out, this is the founding ideology of the Internet, and it has clear similarities with other decentralized systems—most notably the federal system of the United States. Because the early Internet ran on the existing wires of the telephone network, its founders were forced "to invent a protocol that took account of the existence of many networks, over which they had limited power," Wu writes. It was "a system of tolerated difference—a system that recognized and accepted the autonomy of the network's members."

But while this autonomy came about because of the infrastructure the Internet was given, it soon became the crucial force shaping the infrastructure the Internet made. Winston Churchill said about architecture that "we shape our buildings, and afterwards our buildings shape us," and the same is true of the Internet. With TCP/IP in place and new autonomous networks popping up with increasing frequency, the Internet grew physically, but haphazardly. It took shape in ad hoc ways, like a city, with a loose structure giving way to spontaneous, organic growth. The Internet's geography and shape weren't drawn up in some central AT&T engineering office—as the telephone system was—but rather arose out of the independent actions of first hundreds, and later thousands, of networks.

With TCP/IP in place, the Internet—more or less as we know it today—had arrived, and a remarkable period of growth began. In 1982, there were only 15 networks, or "autonomous systems," on the Internet, meaning they communicated with TCP/IP; by 1986 there were more than 400. (In 2011, there were more than 35,000.) The numbers of computers on those networks ballooned even faster. In the fall of 1985, there were 2,000 computers with access to the Internet; by the end of 1987 there were 30,000, and by the end of 1989 there were 159,000. (In 2011, there were 2 billion Internet users, with their hands on even more devices.) The Internet, which had for nearly twenty years been a college town called the ARPANET, had begun to feel more like a metropolis. If before you could imagine each router as a cloister on a quiet mountaintop, the incredible growth in the number of machines meant that those routers were now piling up near one another, forming villages. Some of those villages were even beginning to reveal the vague promise of a skyline. For me, it's the most exciting moment of this early history: the Internet was becoming a place.

Toward the end of the 1980s a handful of companies began to build their own long-distance data highways, or "backbones," and city streets, or "metropolitan" networks. But strike from your mind any images of bulldozers steaming across the Pennsylvania countryside laying cable—although those would come soon enough. These early long-distance and local networks still worked across the existing phone lines, with specialized equipment installed on either end. By the early 1990s, the trickle became a wave, as companies like MCI, PSI, UUNet, MFS, and Sprint attracted increasing investment dollars—and used them to dig their own trenches and fill them with the new technology of fiber optics, which had been commercialized in the 1980s. The network of networks was accumulating an infrastructure of its own. It began to colonize key places around the world—indeed, the places where it still predominantly exists: suburban Virginia and Silicon Valley, California; London's Docklands district; Amsterdam, Frankfurt, and downtown Tokyo's Otemachi district. The Internet had propagated to the point of becoming visible to the naked eye, becoming a real landscape all its own. What for the first twenty years of the Internet's existence were easy to dismiss as in-between spaces—telecom closets and spare classrooms—now had character. By the mid-1990s the wave of construction became a torrent, and "broadband" became one of the most infamous bubbles in American economic history. Yet it was that spending, as overheated and economically destructive as it was, that built the Internet we use today.

In 1994, I was finishing high school, logging long hours on the family Macintosh, endlessly tying up the phone exploring the message boards and chat rooms of America Online. Then sometime that winter my father brought home a small 3.5-inch disk loaded with a new program called "Mosaic"—the first web browser. On a sunny weekend morning, sitting at the dining room table, my physics homework pushed to the side, a long telephone cord strung across the room, we listened as the screeching tones of the modem signaled a connection with a distant computer. My mother looked disapprovingly over the top of her newspaper. On the screen, rather than the America Online menu, with its short list of choices, there was a blinking cursor inside an empty "address bar"—the ur-starting point for all our digital journeys.

But where to go? At that point, the options were limited. Few organizations had websites—only universities, a few computer companies, the National Weather Service. And how did one know where to find them? There was no Google, Yahoo!, MSN, or even Ask Jeeves. Unlike the walled enclosure of AOL, unlike any other computer I'd ever used, it felt as open-ended as the world. The sensation was distinct: it felt like travel. I wasn't alone in the thrill. That was a heady season for the Internet. Netscape released its web browser in October, as Microsoft ramped up the advertising campaign for its own "Internet Explorer." The Internet was about to go mainstream, once and for all. The roof was about to blow off.

But which roof, really? The boom would strain the Internet's existing infrastructure to the breaking point. So who would save it? How did it expand? To where? I'd heard the business stories of the dot-com boom, about how Jim Clark and Marc Andreessen founded Netscape, and Bill Gates battled to keep Internet Explorer an integral part of Microsoft's Windows operating system. But what about the networks themselves, and their places of connections? In a business that's always been obsessed with the next new thing, who was still around to tell that story?

I went back to California—only to hear about Virginia.

On a characteristically damp and gray San Francisco day, I met a network engineer named Steve Feldman at a café a few blocks from his office, in the heart of the cluster of Internet companies located south of Market Street. He looked like a high school math teacher, with khaki pants, sturdy brown walking shoes, and a big beard. His office ID hung around his neck from a lanyard embroidered with NANOG, the North American Network Operators' Group—the clubby association of engineers who manage the biggest Internet networks, and whose steering committee Feldman chairs. These days his job is to run the data network for CBS Interactive, making sure, among other things, that the latest episode of *Survivor* or the NCAA basketball game is properly streamed to your screen. (Even if he wasn't a fan himself.) But for a time in the '90s, Feldman ran the Internet's single most important place, a global crossroads improbably located in the parking garage of an office building in a suburb of Washington, DC. It was a thrilling moment in the Internet's evolution—for a while. By the end, things had gotten out of control.

We sat down between two young guys pecking away at laptops, their heads in the cloud. Our conversation must have sounded strange to them; it was all such ancient history. In 1993, Feldman—a graduate of the computer science department at Berkeley—went to work for a young networking company called MFS Datanet, which had started out laying fiber in Chicago's coal tunnels, and had more recently been building private networks linking corporate offices together, mostly piggybacking on existing phone lines. MFS wasn't providing Internet access itself, only helping companies with their internal networks, but it had gotten good at doing so across a city—which was exactly what the handful of companies that were providing Internet access needed. They had a problem. At the time, the de facto backbone of the Internet was run by the National Science Foundation and known as NSFNET, but technically the commercial companies were prohibited from using it by the "acceptable use policy," which in theory limited traffic to academic or educational purposes. To grow, the commercial providers had to find a way to exchange traffic across their own private roads, in order to avoid traversing that government-run highway. That meant connecting to one another—physically. But where?

Business was booming for everybody, but the whole endeavor was threatened by an absence of real estate. Where could they connect? Quite literally: Where was a cheap place with plenty of electricity where the engineers could string a cable from the router of one network to the router of another?

The Virginia suburbs west of Washington, DC, were already a hot spot for many of the early commercial Internet providers, mainly due to the concentration of military contractors and high-tech companies in the area. "It was the technology center," Feldman told me. For a time, a few of the early Internet providers interconnected their networks inside a Sprint building in northwest Washington, but it was an imperfect solution. Sprint didn't like its competitors setting up shop inside its building (especially when Sprint didn't have a business setup to properly charge them for it). And for the Internet providers themselves—companies like UUNET, PSI, or Netcom—it was expensive to be there, because of the cost of leasing local data lines back to their own office or network POP (or "point of presence").

MFS offered a solution: it would turn its offices into a hub. The company already had plenty of existing local data lines, which it would use to tether each of the Internet providers, like dancers around a maypole. MFS would then provide a switch, called a Catalyst 1200, that would route traffic between the networks. It wasn't merely a local road; it was a roundabout. By plugging into this hub, each network would have immediate and direct access to all the other participating networks, no highway tolls required. But for the plan to be viable a handful of the Internet providers had to commit simultaneously—or else it would be a roundabout in the middle of nowhere. A group of them made the decision over lunch one day in 1992 at the Tortilla Factory in Herndon, Virginia. At the table were Bob Collet, who ran Sprint's network; Marty Schoffstall, cofounder of PSI; and Rick Adams, founder of UUNET (who would later make hundreds of millions of dollars taking it public). Each of these networks operated independently, but they knew full well they were useless without one another. The Internet was still for hobbyists—an eccentric subset of the population, composed mostly of people who had used the network in college and wanted to keep going. (In the United States, the percentage of households with access to the Internet wasn't measured at all until 1997.) But the growth trend was clear: for the good of the Internet—really if there were to *be* a functioning nonacademic Internet at all—the networks had to act as one. MFS called its new hub a "Metropolitan Area Exchange." To indicate its ambition to build several of them around the country, it nicknamed the hub MAE-East.

It took off immediately. "MAE-East was so popular that we were outgrowing the technology faster than we could upgrade it," Feldman said. When a new Internet service provider sprang into existence, its customers would mainly call in over a regular telephone line using a modem. But then the provider had to connect to the rest of the Internet (as Jon Auer did in Milwaukee). And for a time, MAE-East was it. "If you connected to MAE-East, you'd have the entire Internet at your doorstep," Feldman said. "It was the de facto way into the Internet business." Within a couple years, MAE-East was the crossroads for fully half of all the world's Internet's traffic. A message from London to Paris most likely went through MAE-East. A physicist in Tokyo querying a website in Stockholm went through MAE-East—on the fifth floor of 8100 Boone Boulevard in Tysons Corner, Virginia.

It was a portentous location. The intersection of the Leesburg Pike and Chain Bridge Road may have been the crossroads of the digital world, but it was also conspicuously close to the crossroads of American espionage—overhanging MAE-East with an ongoing fog of mystery, suspicion, even conspiracy.

Tysons Corner is one of the highest points in Fairfax County, at five hundred feet above sea level. During the Civil War, the Union army took advantage of its views back toward Washington and out toward the Blue Ridge Mountains, and erected a signal tower there, pillaging timber for its construction from nearby farms. A century later, at the start of the Cold War, the US Army built a radio tower on the same spot and for the same reason: to relay communications between headquarters in the capital and distant military posts. A military tower still stands on the site, a red-and-white steel skeleton looming above a busy suburban crossroads, ringed by a protective fence with a sternly worded sign prohibiting photographs. Heightening the place's mysteriousness, shortwave radio hobbyists have fingered the tower as a source of the "numbers stations"—radio broadcasts of an endless cadence of spoken digits. If the professional spook commentators are to be believed, far-flung spies tune in at specified times to receive coded communiqués from headquarters. According to Mark Stout, a historian at the International Spy Museum, the single-use codebooks the system uses are uncrackable. "You really truly cryptanalytically have no traction getting into a one-time pad system," he says. "None at all."

Indeed, if counterespionage were your gig, the rest of Tysons Corner might pose similar challenges. MAE-East isn't there anymore—or rather, whatever networking equipment that still is there is no longer a significant center of the Internet—but the neighborhood remains the same. Circling the parking lots, the buildings themselves seem sealed, with perfectly flat glass façades, as if conceived by their architects to be as anonymous as they are impenetrable. The buildings are mostly unmarked, in accordance with the wishes of their low-profile tenants. When they do have signs, they reveal the identities of military contractors: Lockheed Martin, Northrop Grumman, BAE. Many were built with special rooms, known as Sensitive Compartmented Information Facilities, or "skiffs," designed to meet government criteria for handling classified information.

The most paranoid network engineers—the "tin-foil hat guys," as they're known, in reference to the conviction that the only way to keep the government from reading your mind is to wear a helmet made of aluminum foil—took MAE-East's location as proof of its malevolent government control. Why else would it be down the road from the CIA? And if it wasn't the CIA listening in, then it must have been the supersecret National Security Agency systematically tapping everything passing through—a claim repeated in James Bamford's bestselling 2008 book about the NSA, *The Shadow Factory*. Even today, do any idle googling about MAE-East and the information seems oddly sketchy—written in the present tense, even though its importance is long past; marked in red on satellite photographs with its relation to a nearby CIA facility highlighted; somehow frozen in time. MAE-East remains an international woman—or an international *something*—of mystery.

Alas, it's all a little overblown. MAE-East's importance may have begun spontaneously, but it continued bureaucratically. In 1991, the US Congress had passed the High Performance Computing and Communication Act, better known as the "Gore Bill," named after its original sponsor, then-senator Al Gore. It's to this that Gore owed his purported claim of having "invented the Internet"—which isn't as far-fetched as it sounds. *Invent* is undoubtedly the wrong word—and Gore never actually said it—but the push from government was crucial in getting the Internet out of its academic ghetto. Among the bill's provisions was a piece of policy best known by its popular name: the "information superhighway." But rather than putting shovels in the ground to build it, government policymakers catalyzed private companies to do it for them, by funding the construction of "on-ramps." A network access point, or NAP, as they called it, would be "a high-speed network or switch to which a number of networks can be connected via routers for the purpose of traffic exchange and interoperation." It would be funded with federal dollars, but operated by a private company. An access point, in other words, would be a network that connects networks: a copycat MAE-East.

Feldman responded to the government request for bids with an idea for a fancy new exchange—but the National Science Foundation, which ran the process, said they'd rather just give MFS money to keep MAE-East going. Contracts were eventually awarded for four access points, run by four major telecom players: the Sprint NAP in Pennsauken, New Jersey, just across the Delaware River from Philadelphia; the Ameritech NAP in Chicago; the Pacific Bell NAP in San Francisco; and MAE-East. But Feldman likes to say there were really only three and half, "because we already existed." (And MFS would soon open MAE-West, at 55 South Market Street, in San Jose, California, to compete with the Pacific Bell NAP.) That geography was deliberate. The National Science Foundation knew that to succeed the network hubs needed to serve distinct regional markets, spread evenly across the country. Distance mattered. The original solicitation accordingly identified "California," Chicago, and New York City as "priority locations." The decision to locate the Sprint NAP in a bunker of a building in Pennsauken, ninety miles from New York, was because of the existing facility's links to the transatlantic undersea cables that landed on the New Jersey shore; it was the gateway to Europe.

The opening of the network access points also marked an important philosophical shift, one that would have ramifications for its physical structure. In a clear departure from its original roots, the Internet was no longer structured as a mesh, but rather was entirely dependent on a handful of centers. As the urban theorist Anthony Townsend has pointed out, "The reengineering of the Internet's topology that was implemented in 1995 was the culmination of a long-term trend away from the idealized distributed network . . . envisioned in the 1960s." As the number of networks increased, their autonomy was best served by centralized meeting points.

But for Feldman the meeting point felt more like a choke point. By 1996, MAE-East was overstuffed with belching, blinking machines, and growing somewhat out of control, however profitably. The original concept had been that each network would house its own router and link into MAE-East over its data lines. A machine evocatively called a FiberMux Magnum would act like the can on one end of a string telephone, changing the signals coming over the line into a form MAE-East's router could understand. But as you might guess, FiberMux Magnums themselves take up space, and the fifth-floor suite at 8100 Boone that housed MAE-East—or you might say that *was* MAE-East—quickly filled up. The situation deteriorated further when networks discovered that they could increase their performance if they chucked the FiberMuxes and installed their actual routers at MAE-East too, in effect making it their new technical office. And it got even more crowded when they discovered performance increased again if they put their servers there as well, so that MAE-East wasn't just the transit point for data, but often its source. Web pages loaded faster for its customers, and it reduced the costs of moving the bits around. But with those changes, MAE-East had transformed from a crossroads to a depot.

It fell to Feldman to find a way to expand. The landlord at 8100 Boone had become impatient with its power-sucking ten-ant, so soon the many-tentacled apparatus moved into a plasterboard enclosure carved out of the basement parking garage of the building across the street, at 1919 Gallows Road. Air-conditioning units surrounded the bare white walls, hard up against the underground parking spots. A generic hardware store ACCESS RESTRICTED sign marked the door. The undisputed capital of the Internet was decidedly humble, the kind of space where you'd expect to find floor-polishing machines and toilet paper stocks, not the spinal cord of a global information network. MAE-East's location in a parking garage may have seemed like something out of a spy movie—the one where an anonymous door in a dingy corridor opens up to reveal a huge, glistening, high-tech lair. But the high-tech lair was a hovel.

It drove Feldman to distraction. When he wasn't selecting and installing new equipment, managing the connections between networks, and trying to figure out what everybody needed, he was apologizing. Traffic had been doubling every year, far outpacing what the router technology could handle, not to mention the real estate. The Internet was clogged. At every meeting of the North American Network Operators' Group, Feldman would be asked to stand up in front of his colleagues and explain why the crossroads of the Internet, his crossroads, was perpetually jammed. It wasn't an easy crowd. "People in the NANOG community say what they're thinking," Feldman said. "And they don't pull punches." At one meeting, ultimately exasperated by the complaints, Feldman taped a paper bull's-eye to his chest before taking the stage. There was no getting around it: the model was broken. The Internet needed a new kind of place.

By 1997, 20 percent of American adults were using the Internet—up from nearly zero a few years before. The Internet had proven its usefulness. But it was unfinished, unrealized. Some of the needed pieces were obvious: there had to be new high-capacity long-distance lines between cities; software tools that would enable "e-commerce" and online videos; and new devices that could connect to the Internet faster and more flexibly. But beneath all that was an unmet mechanical need, an unbuilt room in the Internet's basement: Where could all the networks connect? They came up with the answer down the road, in the heart of Silicon Valley—in a basement, in fact.

3

Only Connect

For a couple of years at the beginning of the millennium—during the quiet time after the Internet bubble burst but before it inflated again—I lived in Menlo Park, California, a supremely tidy suburb in the heart of Silicon Valley. Menlo Park is a place rich in a lot of things, Internet history among them. When Leonard Kleinrock recorded his first "host-to-host" communication—what he likes to call "the first breath of the Internet's life"—the computer on the other end of the line was at the Stanford Research Institute, barely a mile from our apartment. A few blocks past there is the garage where Larry Page and Sergey Brin first housed Google, before they moved into real offices above a Persian rug store in nearby Palo Alto. On the morning of Google's public offering, in August 2004, the crowd at the café on our corner was electrified—not, presumably, because they themselves were getting richer by the moment (although maybe), but because it suddenly made everything seem possible again. Indeed, it was that same summer when Mark Zuckerberg moved his fledgling company, then known as The Facebook, from his dorm room at Harvard to a sublet house in Palo Alto. It wasn't big news at the time—the only person I knew on Facebook then was my sister-in-law, still in college—but it was clear that it made perfect sense. As E. B. White said of New York, this was the place you came if you were willing to be lucky. Just as Wall Street, Broadway, or Sunset Boulevard each contain a dream, so too does this corner of Silicon Valley. Most often, that dream is to build a new piece of the Internet, preferably one worth a billion dollars. (Facebook, by the way, recently moved into a fifty-seven-acre campus, back in Menlo Park.)

An economic geographer would describe all this as a "a business cluster." Silicon Valley's unique combination of talent, expertise, and money has created an atmosphere of astounding innovation—as well as what the local venture capitalist John Doerr once described as the "greatest legal accumulation of wealth in human history." Indeed, this place, perhaps more than anywhere else in the world, oozes with a collective belief in the limitless potential of technology, and that technology's potential to turn into limitless money. The aspiration in the air is palpable.

Yet there seems to be a fundamental irony to all this. Among the computer's greatest contributions to humanity has undoubtedly been its ability to connect people in different places. Perhaps more than any other technology in history, the Internet has made distance less relevant—it's made the world smaller, as the saying goes. As the MIT sociologist Sherry Turkle describes, "A 'place' used to comprise a physical space and the people within it." But the ubiquity of the Internet makes that no longer true. "What is a place if those who are physically present have their attention on the absent?" she wonders. "The Internet is more than old wine in new bottles; now we can always be elsewhere." We feel the consequences of this every day—the disconnection that comes as a consequence of connections, as if in a zero-sum game. And yet that isn't the only truth about the network—especially not in Silicon Valley. Undergirding our ability to be everywhere is a more permanent thicket of connections, both social and technical. We can only talk about *being* connected as a state of mind, because we take the physical connections that allow it as a given.

But the evolution of those connections is very specific and has occurred in very specific places—Palo Alto especially. Whatever alchemy goes on there doesn't, or perhaps can't, happen across a wire. At this intensity, connection is an unabashedly physical process. When I lived there, the faithful who fill the cafés always reminded me of priests in Rome, fingering smartphones rather than rosary beads, but similarly sticking close, for reasons both practical and spiritual, to the center of power. They are all there to connect: the gambling venture capitalists, the Stanford engineers, the lawyers and MBAs, and the start-up junkies who smell the future like bloodhounds. The same is true when we start talking about the actual wires.

Palo Alto is only thirty-five miles from San Francisco, but on the day I drove down it was twenty-five degrees warmer, a dry heat thick with the fragrance of eucalyptus. I was meeting two of those Valley stalwarts for lunch at a café on University Avenue, Palo Alto's main drag. Afterward, we would visit the Palo Alto

Internet Exchange—one of the Internet's most important places of connection, past and present.

Jay Adelson and Eric Troyer sat at a sidewalk table, looking out toward the passing crowd, jovial with a couple beers on a Thursday afternoon. They are old friends, onetime roommates, former colleagues, and among the most knowledgeable two people anywhere about how—and, more important, *where*—Internet networks connect to one another. Troyer calls himself a "recovering network engineer," a moniker that both maintains and deflects his geek cred. With his close-cropped gray-flecked hair and wraparound sunglasses, he gave off a relaxed surfer-dude vibe, like the net-geek version of Anderson Cooper. "ET," as he's known in the networking community, works for Equinix, a company that operates "colocation" facilities around the world.

Adelson hired him there; in fact, Adelson founded Equinix, building it up from a loose concept in 1998 to a billion-dollar publicly traded company, before leaving in 2005. He is the Silicon Valley archetype: an entrepreneur with a gift not only for seeing the future but convincing others to meet him there. He maintained a reputation as a boy wonder, but he was a few weeks shy of his fortieth birthday, and a few months past his most recent job as CEO of Digg, a web service that allows readers to express their approval or disapproval of an online article or blog post, or photograph of a talking cat. Adelson's departure from Digg was widely reported to be contentious, but he seemed relaxed, wearing jeans and an untucked dark button-down shirt, his trademark bangs hanging over an angular face, like a teenager. He had been using his post-Digg sabbatical to take guitar lessons, move into a new $3 million house, spend time with his three kids, and weigh his options for the future—mulling a third act in what had already been a very successful Silicon Valley career. It was the first act that interested me: when he helped solve the problem of MAE-East, and in the process raised the Equinix flag above what are today the Internet's most important choke points.

"You want to hear the whole story?" Adelson said, already gearing up, digging into a chicken Caesar salad. "It was an interesting period, a real transitional point for the Internet." It all happened fast, at the height of the dot-com boom. At the end of 1996, Adelson had a job at Netcom, one of Silicon Valley's first commercial Internet providers. In contrast to the Virginia-based companies focused on big corporate clients, Netcom's bread and butter was "nerds in withdrawal," recent castoffs from university computer science departments desperate to "extend their addiction" to the Internet. Netcom had started out connecting its customers through the academic backbone, even though doing so was in clear violation of its "acceptable use policy." This side door to the Internet had been fine to service the needs of a handful of quiet programmers, but when things took off, it became untenable. So at great expense Netcom leased a data line from its Bay Area headquarters back to Tysons Corner, to join the scrum of networks at MAE-East. Adelson was shocked at what he found there. "It was an old boys club. If you weren't a telecom company, and you weren't controlling the fiber in the ground, you were at a profound disadvantage. They'd tell us, 'We're out of capacity.' But you'd never know if it was a conflict of interest"—if they wanted the business for themselves.

For the Internet to grow, it had to be freed from metered interconnections, carrier interference, and the clogged switches that the National Science Foundation had unwittingly codified with the creation of the network access points. Networks had to be able to connect with as little friction as possible. "We'd post, 'It should be a free Internet! It's unfair for these exchange points to be owned by telecom companies!'" Adelson recalled about the angry debate, played out on the email lists and message boards of the networking community. Because how open was the Internet really if a single company effectively had a velvet rope strung across the door?

Adelson, all of twenty-six at the time, had already distinguished himself as a different kind of networking guy—an Internetworker, you might say. Networking predominantly attracted people who preferred spending time with machines rather than other people. "In order to be proficient in Internet technology at this time, you kind of have to be weird," Adelson explained. He was—a little. He'd been playing obsessively with computers since he was a kid, hanging out on hackers' message boards and spending long hours in the lab in college. But he also studied film at Boston University and had acquired the wheeling-dealing, fast-talking poise of a Hollywood producer. His skill was getting people to connect—as well as computers.

The Internetworking world is still surprisingly small, but it was tiny then, and Adelson attracted the attention of an engineer named Brian Reid at Digital Equipment Corporation, one of Silicon Valley's oldest and most venerable computer companies (now part of the giant Hewlett-Packard—another company born and headquartered in Palo Alto). Digital had a node on the ARPANET nearly from the beginning, but it wasn't until 1991 that it began hosting a crucial private Internet link—a wire strung across the room connecting two of the pre-MAE-East era's biggest regional networks, Alternet and BARRnet. It was originally set up in a spirit of community service—"for the good of the Internet," as the engineers like to say. But as the Internet grew, Digital recognized another benefit: the link gave them a front-row seat to a key Internet intersection. They were like traffic experts with offices overlooking Times Square. And it was getting messy out there.

Digital was particularly sensitive to the failures of MAE-East, because it designed and manufactured the "GIGAswitch/FDDI," the router at its heart that couldn't keep up with demand. To continue to grow, there needed to be a new way for networks to connect to one another that eliminated the congestion problem. Reid had the simple idea that networks should connect directly, literally plugging one router into another, rather than all plugging into a single shared machine as at MAE-East and the other network access points. Most of the networks had already moved lots of equipment into the buildings, which were overflowing because of it. They needed a better environment—a more suitable piece of real estate than a concrete bunker carved out of a parking garage—and one that could accommodate all the direct interconnections. Reid also imagined that the revenue model would change: the connections would be "untariffed," meaning Digital wouldn't charge by the amount of traffic. Instead it would charge rent, both for the obvious square footage of the "cage" in which customers kept their equipment, but also for the far subtler (and skinnier) piece of air that each wire traversed to connect to another company's cage. At MAE-East this would have been commercial suicide, like a restaurant giving away food; but Digital thought it could make a business charging for the table. And the risk was worth it, particularly if it would help grow the Internet—and sell more of Digital's machines.

Digital put in a few million dollars of internal funding and a spare bit of office space: the basement of 529 Bryant Street, constructed in the 1920s as a telephone switching office. In technical terms it would be "carrier neutral," meaning Digital wouldn't be competing with its customers, as at the network access points. And it would be built within a "class A data center," a space specially designed for computer and networking equipment. Reid christened it the "Palo Alto Internet Exchange," or PAIX. All he needed then was somebody to run the place—somebody who knew networking. Somebody with some vision.

For Adelson, Digital's job offer was a funny thing. "I remember thinking, 'Digital?!'" he recalled. "All my friends were going to dot-coms and they were going to make millions as 50 percent owners of their start-ups, and I'm being recruited by a thirty-year-old company! But I was an Internet nerd, and Digital had this nerd cred." Besides, that wasn't exactly how it worked out.

The exterior of 529 Bryant Street was impeccably maintained, its walls a match for the sandstone structures on Stanford's famous quad. Elaborate bas relief engravings surrounded the entrance, as if it were some lost London bank. Brass letters spelled out "PAIX" beside the door. We stepped into the small lobby, hidden from the street by tinted windows. "Oh. My. God," Adelson said, when his eyes adjusted to the relative dark. "That is really, really cool." In front of him was a big red and black "E"— Equinix's logo. Adelson hadn't been inside this building since the bad day in 1998 when the handshake deal that was going to make this the nascent Equinix's first location fell apart, and the place was sold out from under them for $75 million. But only a few weeks before our visit, a dozen years later, Equinix—without Adelson—had finally gotten the PAIX, as a spoil of its purchase of a key competitor, Switch & Data, for $689 million in cash and stock. For Adelson, the Equinix logo on the wall was the symbol of an earlier wrong finally righted—and proof that his vision for the Internet was correct.

Two technicians greeted us, both of whom had worked in the building since those early days. In Internet time, multiple epochs had passed; but amid the hugs and back slaps the span of time felt reassuringly human. The newborn babies were still hardly grown. "I was going to ask what you been doing these last ten years, but of course I know!" Felix Reyes, one of the techs, said to Adelson. "It's good to see you! A lot has changed here, a lot of corporate politics, a lot of growth. But still kicking it!" That seemed like an understatement: the building was on its fourth owner; the Internet was transformed, and had transformed everything.

"It is a long time ago in Internet years," Adelson said.

Reyes was wearing a brand-new Equinix polo shirt, black with the red logo, and Adelson flicked at it. "I don't have any schwag!" he complained. "I always had the technicians' staff shirts."

"We'll get you one," Reyes said. "We have samples of all the shirts over the years."

"This place has changed hands so many times, and morphed again and again, but the reality is the same exact service has been taking place here since its inception," Troyer said.

We headed down a staircase behind the security desk toward the basement, where the first equipment was installed, early in 1997. By the end of that year, the Palo Alto Internet Exchange had grown to be the most important building of its kind on the planet. It no longer claimed that title, but it still remains high on a short list of the Internet's most important places: a key nodal point where networks connect to one another. The building in Milwaukee was the Internet equivalent of a small regional airport, with just one or two airlines flying to a couple big regional hubs; but the Palo Alto Internet Exchange is like San Francisco International, or even bigger—a "major global connectivity hub," in the words of Rich Miller, a key industry observer. The building all around us was the aggregate manifestation of those connections. It provides the real estate to satisfy a basic economic and technical desire: it is cheaper and easier to connect two networks directly than to rely on a third network to do it for you. The PAIX is the depot: a convenient central point to string a cable from one router to another. And in particular, it's a popular place for the undersea cables linking Asia and North America to install their network POPs, or "points of presence." This is the place that makes "connect" a physical word.

From a modest foyer at the foot of the stairs, I could see tight rows of cages stretching far off into the half-light, like the stacks of a library. Each was about the size of a cubicle and rented by a single network, which installs its equipment and starts arranging connections to other networks—looking to literally string a wire. Originally, the companies that owned the long-distance fiber-optic lines came to the building to be close to the local and regional Internet service providers that brought the Internet to homes and business—the "eyeball" networks, they're called. These were the physical network owners. But soon the "content providers"—a Facebook or YouTube today, but back then a Yahoo!, an electronic greeting card company, or a pornography site—wanted to be close as well, to improve the connections to their eyeballs. "I remember Filo and Yang from Yahoo! coming through here, and thinking, 'Who are these clowns?'" Adelson said about Yahoo!'s billionaire cofounders. But as the Internet evolved, eventually everybody showed up, from virtually everywhere, over a hundred networks in all. Today, there are big content players like Microsoft, Facebook, and Google; eyeballs like Cox, AT&T, Verizon, and Time Warner; and then the global telecoms large and small, with a particularly strong showing from the Pacific Rim—everyone from Singapore Telecommunications to Swisscom to Telecom New Zealand to Qatar Telecom to Bell Canada, coming in on the transpacific cables or the big backbones crossing the United States. Like a throbbing world capital, the PAIX thrived on its own diversity.

As we stepped into the dimly lit corridor lined with cages, in front of us was an enormous cardboard box the size of a shower stall. Inside was a brand-new router, the most powerful model made by Cisco, one of the industry leaders, with a six-figure base price. Only the largest websites, corporations, or telecom carriers would have enough traffic to justify a beast like this. Finding it sitting there waiting to be set up was like seeing a brand-new 747 parked idle on the airport tarmac. But what made it special wasn't just the volume of data it could move but the number of different directions it could move it. In that sense, to shift the analogy, the big router was more like the traffic roundabout at the meeting point of 160 highways—160 being the number of individual "ports" it accommodated, each with a processor that handled the communication with another router, like a two-way street. It was vastly more powerful than the old Catalyst or GIGAswitch boxes used in Tysons Corner. But even more remarkably, it represented the needs of just one network, rather than being at the center of many. It wasn't the singular machine at the building's heart, but one of hundreds all connected to one another.

Those connections are always physical and social, made of wires and relationships. They depend on the human network of network engineers. Early in his career, Troyer spent his fair share of time sitting on the floor of one of these cages, wrestling with a broken router. But more recently his job was to be more of a social director, encouraging networks to connect with each other—with Equinix collecting a monthly fee when they did. What surprised me was how personal that process was. Troyer knew the network engineers; he was friends with them on Facebook and made sure to buy them beers. The Internet is built on connections between networks agreed on with a handshake and consummated with the plugging in of a yellow fiber-optic cable. Technically the connections happening here could happen across any distance. But there's a profound efficiency in doing it directly, in plugging my box into your box, in an exponentially repeating pattern.

Walking through the PAIX is a lesson in the "network effect," the phenomenon by which something becomes vastly more useful the more people use it—leading more people to use it. In Palo Alto, the more and larger Internet players moved into the building, the more and larger players wanted to be there, seemingly ad infinitum, up against the laws of physics—and the Palo Alto city council. All this equipment needs backup electrical generators, in case of a blackout; and generators need vast quantities of stored diesel fuel—more than any of the neighbors would prefer. "We ran this facility right to the max of what it could handle," Adelson recalled.

As we walked through the dim corridor between cages, the physical ramifications of all those connections were above us: thick rivers of cables in bundles the size of tires, laid in racks suspended from the ceiling, and then cascading down in "waterfalls," as the techs call them, into each cage. The building buzzed with their energy. "You're being irradiated as you speak," Troyer said, only half-jokingly. "Jay's already had three kids so he's okay." There were upward of ten thousand Internetwork connections, or "cross-connects," in this building alone. This was the tangle of dusty cables hidden behind my couch inflated to the scale of a building—and it wasn't any easier to organize.

In the PAIX's early days, "cable management" was a crucial technical challenge. The Internet was tangled. Experimenting with different ways of handling things, at one point Adelson and his staff tried prewiring different zones of the building to create fixed paths that could be patched together as needed, like an old-fashioned telephone switchboard. "But what we learned—or what poor Felix here learned—is that every time you did that you introduced a point of failure," Adelson recalled, while Reyes shook his head at the memory. So they stuck to laying cable on an as-needed basis. A few years later, one particularly talented cable layer named John Pedro would earn US patent 6,515,224 for his technique: a "cascading cable tray system" with "pre-fabricated support structure."

As we walked between the cages filled with boxes glittering with blinking green lights, I had to remind myself to try to associate what I was seeing with its effect in the real world, in people's lives—to confront, in the most basic terms, the way things move across the Internet. That required a leap of the imagination. Let's say that yellow wire there belonged to eBay: Whose jade collectible teapot was zipping across it? Or what did a winemaker in New Zealand have to say to a sheikh in Qatar? My phone was on and getting emails. Were they passing through here? My niece lost a tooth—was the picture passing through Facebook's cage over there?

But the Internet as it surrounded me wasn't a river I could dip a net in and pull up a sample to count the fish. To find the scale of information as we experience it each day—to find, say, a single email—would be more akin to counting the molecules of water. Each of those fiber-optic cables represented up to ten gigabits of traffic per second—enough to transmit ten thousand family pictures *per second*. The big router had up to 160 of those plugged in at once; and this building was filled with *hundreds* of those routers. Walking through the dimly lit aisles was like hacking through an underbrush of quadrillions, an unfathomable quantity of information.

Yet for Adelson, there was a time when it was all personal. He saw a story in every corner. "Remember when we shut off Australia?" he excitedly asked the ad hoc tour group, stopping in front of one cage, a little emptier than the others. A router for

the Australia Internet Exchange—"Ozzienet or something"—
was installed in the building, but they weren't paying their bills.
Adelson remembers the phone call he received at home the evening they finally pulled the plug: "My wife was, like, 'There's
somebody on the phone, they're not happy, and it's something
about the Internet in Australia being down?' I was, like, 'Really?
Give me that.'"

In another cage was the onetime home of Danni's Hard Drive,
a prominent early pornography site—online home of Danni
Ashe, who *Guinness World Records* once named the "Most
Downloaded Woman" (a category they no longer track). One
night in the late '90s, Danni herself was purportedly discovered
here in the basement, naked with her eponymous hard drive, in
the midst of taking the "photo of the week." The old-timers nodded at the memory, but later I'd hear the same legend repeated
at other big Internet buildings, and when I eventually tracked
down Ashe and her network engineer at the time, Anne Petrie,
they placed the event not in Palo Alto but at MAE-West, the
Silicon Valley cousin of MAE-East. "I am the woman formerly
known as Danni Ashe," she wrote to me. "Unfortunately, I don't
remember a whole lot of details from that day but I'm guessing
the two engineers working for me at the time would." Indeed,
Petrie did. She had spent sixteen hours installing a pair of new
SGI Origin servers, the state of the art at the time, and Ashe and
her husband had come to see them. "Invariably whenever Danni
was on television everything would go down, because it would
just flood the servers with requests," Petrie recalled. The photo
shoot commemorated the occasion.

We worked our way deeper into the building and back in

time. Adelson stopped in front of a cage that was a bit larger than the others and pleaded with the guys, "Can we go in here? I won't touch anything. I gotta go in here!" The space was more like a small office than a cubicle and was built into the corner of the building, so that two of its walls were solid, rather than the typical steel mesh. It was filled with ancient-looking equipment, pocked with little steel toggle switches and an old black telephone headset.

"This is where one of the most important lies I've ever told took place," Adelson announced, with mock gravity. From the beginning, the PAIX was "carrier neutral"—but at the beginning it was also "carrier free." It was like that week after moving into a new home, before the cable guy arrived. It wasn't connected. One of Adelson's biggest challenges was convincing competing fiber-optic network owners to "pop" the building and establish a "point of presence"—a place to connect. But at the time the carriers just didn't do it. They kept their own equipment in their own facilities and you came to them, paying an arm and a leg for the "local loop" required to do so. (This was the situation out of which MAE-East was born: its parent company, MFS, was in the local loop business, and MAE-East was essentially a very local loop.) If one carrier came into the PAIX, they knew others would. So Adelson lied. "I went to Worldcom and said, 'Pacific Bell said they're going to come in in about three weeks.' And they're, like, 'Really?'" Then Adelson went to Pacific Bell (now AT&T) and said the same thing: "Guess who's installing their fiber backbone in the basement . . . ?" They panicked—their local loop monopoly was at stake. "They're, like, 'We're coming in!' We said we had orders—but we made the whole thing up."

Adelson pointed up at the ceiling, where a thick bundle of black cables disappeared into a dark hole. It was the sort of business decision—like bringing a hot dog vendor to a ball game—that one only needed to be convinced to try once; ever after, it would be recalled only as the smartest thing they'd ever done. After that, the building filled up so fast it became a struggle to keep up. Every available inch was taken over for equipment.

"There were racks in the bathrooms!" Adelson said, as we continued upstairs to see what was once office space and now has been entirely converted for networking equipment. "How many times did we get to a point—just in the first two years— where we said, 'It is not physically possible to do anything else in that building,' and then a month later, be, like, 'Okay, we found a way!'" They called the place "the Winchester Mystery House of Internet buildings," an allusion to the haunted Silicon Valley mansion of the Winchester rifle heir, who for thirty-eight years obsessively added rooms in a desperate effort to evade the ghosts she believed her fortune had created. The PAIX was similarly an exercise in creative construction. Wedged as it was into a downtown Palo Alto block, there was no room for horizontal expansion. The city government had little appetite for the ever-increasing quantities of fuel needed to run the backup generators. Seismically, the structure was just barely up to snuff—for office space, not heavy computer equipment. Adelson shook his head at the memory. "You couldn't have chosen a worse building."

But the Palo Alto Internet Exchange's real trouble came from a different direction. At nearly the same moment the building became the Internet's dominant switching point, in January

1998, Digital, its parent company, was acquired by the Compaq Corporation for $9.6 billion, in what was at the time the largest deal in the history of the computing industry. It was bad news on Bryant Street. As Compaq and Digital struggled to integrate, there was growing concern that the relatively small business of the PAIX would fall through the cracks—and at just the moment when the Internet needed it most. The PAIX had quickly set the standard for how, under a single roof, the network of networks that compose the Internet connect. But the PAIX's success was also its Achilles heel: there wasn't enough of it. The PAIX made it clear that carrier-neutral exchanges worked. But it also made it clear that they needed room to breathe—that an old building in a dense (and expensive) downtown wasn't ideal.

Adelson saw an opportunity. This was the moment when everybody and his mother had an idea for a "dot-com," typically one that would use the Internet's virtualizing power to transform an industry: from grocery delivery to auctions, movie listings to classified ads. But if most people saw the Internet as a means to leave the physical world behind by setting up virtual storefronts or auction halls, Adelson saw an unmet need in the opposite idea: all that virtual stuff needed a physical world to call its own.

There would be Palo Alto Internet Exchanges everywhere. Adelson would be like the Conrad Hilton of the Internet, opening up a branded chain of "telecom hotels," where network engineers could be assured of a consistent experience. Unlike the buildings owned by the big telecom carriers themselves—like Verizon or MCI—these would be neutral places where competing networks of all kinds could connect. Unlike MAE-East or, to a lesser degree, the PAIX, they would be built with the proper backup and security systems, and designed to make it as easy as possible for networks to connect with one another. And like a free morning *Wall Street Journal* at a business hotel, the exchanges would offer perks meant to specially appeal to their unique customers: the network engineers (and former network engineers), like Adelson himself. The challenge was figuring out where on the big planet Earth to put these places. How many of them did the Internet really need?

Adelson relied on a crucial hunch about how the structure of the Internet would evolve: networks would need to interconnect at multiple scales. They had to not only occupy the same building but the same building in several different places around the world. The networks of the Internet would be global, but the infrastructure would always be local. In that sense, the Hilton analogy doesn't require stretching at all: Equinix wasn't trying to establish a single central point, but a short list of capitals in the most important markets—mirroring the tendency of big multinational corporations to have offices in the same short list of global cities, from New York to London to Singapore to Frankfurt. An Equinix "Internet Business Exchange" would be the same place everywhere. For me—a traveler to the Internet—this presented something of a paradox: Was an Equinix facility best understood as locally distinct and unique, or just one part of a continuous global realm, a wormhole across continents? Was an Equinix data center a place or placeless? Or both?

When Adelson left Digital, he and a colleague, Al Avery, quickly raised $12.5 million in venture capital, primarily from some big names with a vested interest in the Internet's growth, including Microsoft and Cisco, the router company. The "ix" in

Equinix indicated an "Internet exchange"; the "equi," their intent of being neutral and not competing with their customers. According to the arrangement made before Adelson left Digital, Equinix was to acquire the PAIX from Digital as its first location. But that never happened. In a turn of events that Adelson always thought of as a betrayal, the private deal turned into an open auction, and the PAIX slipped through his fingers.

But the loss didn't just change the fledgling Equinix's business strategy, it indelibly shifted the geography of the Internet. On the assumption that Equinix had the West Coast covered, at least to start, Adelson focused his attention on Virginia, where MAE-East was still the traffic-snarled center. There were basic macro advantages to this move. The sheer geographic size of North America made it inefficient to send data back and forth across the country, especially multiple times. The fifty-millisecond trips added up, noticeably slowing things down. Compounding the problem, most intra-European Internet traffic was coming to the United States to move between networks; the regional centers were still in the future. The four thousand miles from Paris to Washington was far enough, without adding another twenty-five hundred across the continent. The East Coast needed a hub—and a more efficient one than MAE-East.

Zooming in, Adelson saw that the obvious way to compete would have been to put a new Equinix building right into Tysons Corner. But that wasn't an option. The place "had been in telco hell for too long already," Adelson recalled. The surrounding roads had been dug up so many times that Fairfax County planning officials were sick of it. But Loudoun County, farther from DC, was still mostly farmland in the shadow of Dulles airport. And county officials wanted in on the action. Adelson remembers the big poster in the lobby of the county office showing a fistful of telephone cables backlit with purple light and the hopeful slogan "Where the fiber is." Fiber was what Equinix needed—lots of it, and from multiple carriers, as at the PAIX. It was the sunlight in the greenhouse. Loudoun County officials were eager to help the company get it, even going so far as to offer Equinix the rights-of-way needed to "trench in"—literally dig a hole—to the front door of the building. And this time Adelson knew he wouldn't have to lie to the carriers to get it. The PAIX had quickly become a gold mine for them, and Equinix was offering the same formula but on a larger scale. The timing couldn't have been better. The broadband rush was on, with billions of dollars of investment being made to build multiple major new nationwide fiber-optic networks.

For help selecting a site, Adelson hired a construction company fresh off one of those fiber builds and had its employees bring their maps along with them to the job. Together, they zeroed in on a small parcel of land wedged between Waxpool Road and the disused Washington & Old Dominion railroad tracks, about three miles from the tip of Dulles's runways, in the unincorporated town of Ashburn. The fledgling Equinix bought the land outright. It had to own the dirt. The building Adelson had in mind could not be moved down the block a few years on. Once in place, it would be a delicate and immovable ecosystem, like a coral reef formed out of the steady accretion of networks.

But at the time the place was just empty—or at least that's how it felt to Adelson after the constraints of Palo Alto. The PAIX had been limited (and still is to this day) by its location downtown.

But Ashburn amounted to a declaration of the Internet's manifest destiny. The network of networks would no longer be beholden to legacy telephone infrastructure wedged into crowded cities. Instead, the Internet could expand into America's virgin countryside, where the room for growth appeared limitless.

———————————

Today, Ashburn, Virginia, is a small town that Internet people think of as a giant city. They toss around "Ashburn" as if it were London or Tokyo, and often in the same sentence. Equinix's unmarked complex sits behind an Embassy Suites hotel, no larger or any less nondescript than the small warehouses and light industrial buildings up and down the block. On the hot June day I first visited, a maintenance worker wearing a surgical mask swept the empty sidewalk. Jetliners buzzed by low overhead. Heavy-duty power lines hemmed in the horizon. The surrounding neighborhood was so new that when I tried to drive around the block, the unblemished blacktop streets soon gave way to gravel. My GPS showed me to be crossing open ground. The map hadn't caught up to the sprawl. On either side of the road, industrial-sized driveways led fifty feet into lush green meadows and then stopped, as if awaiting further instructions.

On another visit a year later things had changed: the Embassy Suites was still there, along with the Christian Fellowship Church in a big-box building next door, like a Home Depot. But the empty meadows on the far side of the tight little Equinix campus had been filled in with what looked like two beached aircraft carriers. These were massive new data centers built by a competitor, DuPont Fabros, in an explicitly parasitic arrangement—a

Burger King across the street from a McDonald's. This was a clue as to Ashburn's particular importance. It was the extreme logical opposite of the Internet's standard proposition, and the evolution of Adelson's initial insight: if most days we count on the Internet to let us be anywhere, Ashburn had indeed become an utterly unique place on earth—a place worthy of pilgrimage.

When I showed up, I had trouble finding the door. Equinix had grown to fill six single-story buildings at the time I visited; by early 2012, four more had been added, totaling more than seven hundred thousand square feet—about the size of a twenty-story office building—all tightly arranged around a narrow parking lot. I saw no proper entrance to speak of and no signs, only blank steel doors that looked like fire exits. But the parking lot was full, and I followed a guy into the security lobby of what turned out to be the wrong building. When I finally found Dave Morgan, the director of operations for the complex, he saw no reason to apologize. On the contrary, confusion was his goal: customers are reassured by the anonymity of the place "except maybe on their first visit." Then he shared a handy tip for the next time I found myself similarly lost on the way to the Internet: look for the door with the ashtray next to it.

The lobby was dramatically lit with halogen spotlights. There was executive waiting room furniture, a pair of uniformed security guards nestled behind bulletproof glass, and a big TV tuned to CNN. Troyer was waiting inside. He'd flown in from California to give a proper tour. He had experienced this place in his previous job as a network technician at Cablevision, a New York–area cable company (and owner of my squirrel-chewed wires). Cablevision had always been ahead of the curve offering high speeds to its customers, which meant it had a lot of Internet traffic to move. It was Troyer's task to move it as efficiently—and cheaply—as possible. He extended Cablevision's backbone here from New York, to connect directly to as many different networks, and thereby reduce the amount the company was paying the middlemen, or "transit" networks, to do it for them. It was cheaper for Cablevision to lease its own "pipe" all the way to Virginia than to depend exclusively on the local options in New York (of all places). The Internet's geography is indeed particular. (Not that there was any trench digging involved; the company merely leased capacity on an existing fiber pathway.) "That's the point of view for most major network service providers," Troyer explained. " 'Where can I send my network data physically—geographically—in order to get the most vectors?' Or, to go back to the pipe analogy, 'Where can I drag my data to where there are the most pipes available to send them on the shortest path possible?' " This was the identical issue faced by the guys at the Tortilla Factory (just down the road from here) when they decided to move into MAE-East. And it was the identical issue Adelson faced while at Netcom. For all of us sitting in front of our screens, the Internet only works because every network is connected, somehow, to every other. So where do those connections physically happen? More than most anywhere else, the answer is "Ashburn."

At Equinix, Troyer's job was connecting—in a social way—with people (like his former self) who manage large Internet networks and are always looking for more places to bring them, without relying on any new middlemen. It suited him, as something of an extrovert in a crowd of introverts. He'd be right at home selling television time or mutual funds, or something else similarly abstract and expensive. But occasionally he'll switch modes from jocular salesman to net-geek, offering a soliloquy of technical protocols and operating specifications, his jaw clenching with the effort of precision. Even the most social of networking guys know the geek that lies within. No doubt he isn't alone at Equinix's Silicon Valley offices, which he commutes to from his home in San Francisco. For a period of time after Adelson left the company, he commuted to his job at Digg from New York, sharing a crash pad with Troyer in the Mission District. The Internet is a small world.

And—or so it seemed that morning in Ashburn—a secure one. Getting inside Ashburn required an elaborate identification process. Morgan, the director of operations, had previously registered a "ticket" for my visit in his system, which the guards behind bulletproof glass then checked against my driver's license. Morgan then punched a code into a keypad beside a metal door and placed his hand on a biometric scanner, which looked like an air dryer in an airport bathroom. The scanner confirmed his hand belonged to him, and the electronic lock clicked open. The three of us shuffled into an elevator-sized vestibule—affectionately called the "man-trap"—and the door locked behind us. This was a favorite feature of Adelson's, dating back to the initial Equinix vision. "If I'm going to close that deal with a Japanese telco, and I need to impress them, I need to be able to take a tour group of twenty people through this building," was how he explained it. And it better look "cyberrific," to use Adelson's favorite term. The man-trap wasn't only to control access into and out of the building but also (it seemed) to induce a moment of frisson. For a few long seconds the three of us looked up at the surveillance camera mounted in a high corner, flashing tight-lipped smiles to the hidden guards. I took a moment to admire the walls, hung with illuminated wavy blue glass panels made by an artist in Australia. All the early Equinix data centers have the same ones. Then, after that long, dramatic pause, the airlock doors opened with a hearty *click* and a *whoosh,* and we were released into the inner lobby.

It too was cyberrific. Its high ceiling was painted black, like a theater, and disappeared in darkness. Spotlights left small pools of illumination on the floor. "It's a bit like Vegas," Troyer said, "no day or night." Inside there was a kitchen, snack machines, a bank of arcade-style video games, and a long counter, like at an airport business center, with power and Ethernet plugs where engineers could set up shop for the day. The stools were nearly all occupied. Many customers ship their equipment ahead and pay for Equinix's "smart hands" service to have it "racked and stacked," as they say. But then there are those guys affectionately known as "server huggers," who either by choice or necessity spend their days here. "They're locals, like Norm in *Cheers,* pulling up his bar stool," Troyer said, nodding in the direction of a large guy in jeans and a black T-shirt, hunched over his laptop. "But this is not a resort destination." Beside the kitchen area was a glass-enclosed conference room with Aeron chairs and red speakerphone buttons embedded in the table. A group of half a dozen men and women in business suits had spread out files and laptops on the boardroom table, hard at work auditing the building's systems for a customer, likely a bank. Beside the conference room was a curved, fire-engine red wall. They call it "the silo," a name that evokes ICBMs more than grain. It was Equinix's trademark architectural feature—its golden arches.

Adelson loved that idea: that an engineer responsible for a global network would feel at home in Equinix facilities everywhere. There are about one hundred Equinix locations around the world and all of them carefully adhere to brand standards, the better to be easily navigable by those nomads in endless global pursuit of their bits. Ostensibly, Equinix rents space to house machines, not people; but Adelson's strikingly humanist insight was that the people still matter more. An Equinix building is designed for machines, but the customer is a person, and a particular kind of person at that. Accordingly, an Equinix data center is designed to look the way a data center *should* look, only more so: like something out of *The Matrix*. "If you brought a sophisticated customer into the data center and they saw how clean and pretty the place looked—and slick and *cyberrific* and awesome—it closed deals," said Adelson.

Troyer, Morgan, and I passed through a gate in a steel mesh wall, and it was like stepping into a machine, all rush and whir. Data centers are kept cold to compensate for the incredible heat emitted by the equipment that fills them. And they're noisy, as the sound of the fans used to push around the cold air combines into a single deafening roar, as loud as a rushing highway. We faced a long aisle lined with darkened steel mesh cages, each with a handprint scanner by the door—similar to the PAIX, but far more theatrical. The blue spotlights created a repeating pattern of soft glowing orbs. Everyone at Equinix confesses to aiming for some visual drama, but they're also quick to point out that the lighting scheme has a functional purpose as well: the mesh cages allow air to circulate more freely than in small enclosed rooms, while the dim spot lighting assures a level of privacy—preventing competitors from getting a close look at what equipment you're running.

The Equinix buildings in Ashburn (and everywhere really, but especially here) aren't filled with rows and rows of servers, themselves filled with enormous hard drives storing web pages and videos. They're mostly occupied by networking equipment: machines in the exclusive business of negotiating with other machines. A company like Facebook, eBay, or a large bank will have its own big storage data center—perhaps renting space next door, inside those DuPont Fabros aircraft carriers, or in a building all its own hundreds of miles away, where electric power is cheap, and there's enough fiber in the ground to keep the company connected. Then a company will "tether in" here, running a fiber-optic connection to this distribution depot, and spray its data out from a single cage. (This is exactly what Facebook does—in multiple Equinix facilities around the world, including Palo Alto.) The heavy-duty storage happens in the boonies, in the warehouse, while the wheeling and dealing—the actual exchange of bits—happens here, in the Internet's version of a city, deep in our version of the suburbs, where hundreds of networks have their offices (or cages) cheek by jowl.

I could see the physical embodiment of all those connections above us, where rivers of cable obscured the ceiling. When two customers want to connect to each other, they'll request a "cross-connect," and an Equinix technician will climb a ladder and unspool a yellow fiber-optic cable from one cage to the other. With that connection in place, the two networks will have eliminated a "hop" between them, making the passage of data between them cheaper and more efficient. For the Equinix technicians, laying cables is something of an art form, with each individual type placed at a certain layer, like a data center mille-feuille. Closest to our heads were yellow plastic raceways, the size and shape of rain gutters, typically made by a company named ADC. They come in an erector-set system of straightaways, "downspouts," and connectors, sold at varying widths, depending on how many cables you need to run. The "4 x 6 System," for example, can hold up to 120 3 mm yellow patch cables, while the "4 x 12 System" can manage 2,400 of them. Equinix buys the raceways in such quantities that the company occasionally requests custom colors—clear plastic, or red, in contrast with the stock yellow. The oldest cables are on the bottom of the pile. "It's almost like an ice core," Troyer said. "As you dig down you're going to see sediment from certain time periods." Given the monthly fee charged for each "cross-connect," this is the bread and butter of Equinix's business. The bean counters see each one as monthly recurring revenue. The network engineers see vectors. The data center techs see the sore back they'll get reaching up to the ladder racks to run the cables. But in the most tangible way possible, these cables are the *Inter* in Internet: the space in between.

One layer nearer the ceiling, above the yellow fiber, is the "whalebone," a more open style of cable organizer that indeed looks like the rib cage of a large seagoing mammal. It holds the copper data cables that are physically thicker, stronger, and cheaper than the yellow fiber-optic ones but can carry much less data. Above that is stainless-steel racking for AC power; then a thick black metal frame for DC power; then thick green electrical grounding cables, each layer visible above the other like branches in a forest. Finally, up top near the black ceiling is the "innerduct": ridged plastic tubing through which run the thick fiber-optic ribbons operated by the telecom carriers themselves. This was where Verizon, Level 3, or Sprint would have its cables. Unlike the yellow patch cords that each contain one strand of fiber, the innerduct might have as many as 864 fibers, tightly bonded together to save space. This is Ashburn's spinal cord—the stuff that Adelson fought to bring into the building in the first place—and it accordingly occupies the safest place, out of harm's way near the ceiling. "That stuff's important for us," Troyer said. "Bundled fiber coming in is what gives us our value."

We followed the path of the innerduct to the center of the building, an area known as "Carrier Row." Concentrating the big guys in the middle is practical: it limits the likelihood of having to make a six-hundred-foot run from one corner of the building to the other. But it's also symbolic: these are the popular kids standing in the center of the party, with everybody on the edge craning to see.

We came to a cage with its lights on. Troyer is professionally discreet about what companies have equipment here, but he happily talked through the anatomy of a typical installation. In the near corner of the parking-spot-sized space was the "DMARC," short for "demarcation point," an old telecom term to describe the place where the phone company–owned equipment ended and the customer's begins. It worked the same way here. The heavy-duty metal and plastic rack, the size of a circuit breaker in a large house, was the physical switchboard where Equinix handed off communications cables to the customers. It was the industrial-sized equivalent of a telephone jack: a passive, or "dumb," device, a solid object whose job was to keep the cables neat and organized and make it easy to plug them in. From the DMARC, the cables then ran through overhead trays to the main bank of racks.

Data center racks are always 19 inches wide—a dimension so standardized that it's a unit in and of itself: a "rack unit" or "RU" measures 19 inches wide by 1.75 inches tall. The heart of the operation here was a pair of Juniper T640 routers, clothes-dryer-sized machines designed to send massive quantities of data packets off toward their destinations. These two were likely set up so that if one failed, the other would immediately step in to pick up the slack. Troyer counted the 10-gigabit ports on one of them, each with a blinking green light and a yellow cable sprouting from it. There were seventeen of them. Working together, they could move a maximum of 170 gigabits per second of data—the kind of traffic a regional cable company like Cablevision might put up, serving the aggregate needs of its three million customers. Serious computational power was required to make the innumerable logical decisions to send so many pieces of data out the correct door, having checked them against an internal list of possibilities. That power in turn generated serious heat, which in turn required hair-flattening fan levels to keep the machine from cooking itself. The machines roared with the effort. We all squinted our eyes as the hot air blew at us through the mesh cage wall.

Beside the big routers was a rack holding a couple single-RU servers. These were too small to be doing any serious work actually "serving" web pages or videos. Most likely they were merely running software to monitor this network's traffic—like robotic techs in lab coats, scratching notes on a clipboard. Below those servers was some "out of band" equipment, meaning it was connected to the rest of the world via an entirely separate path than the main routers—perhaps even by an old telephone modem, or occasionally through a mobile data connection, like a cell phone, or both. This was the fail-safe. If something went horribly wrong with the Internet (or more likely just this piece of it), this network's minders could telephone into the big Junipers to fix it, or at least try. You never want to rely on your own broken lines. They always had another option, though: the bulky power strip on the floor, which sprouted not only big electrical cords but also an Ethernet cable connecting it back into the network. Just as you might unplug and plug back in your connection at home to get it working, this did that, but remotely. An engineer could turn the power on and off at a distance—never a bad troubleshooting tip, even for these half-million-dollar boxes. But it doesn't always work. Sometimes the techs had to show up in the flesh to yank the plug.

Equinix Ashburn sees upward of twelve hundred visitors a week, but I wouldn't have guessed that walking around. The size of the place, the twenty-four-hour clock, and the nocturnal predilections of network engineers made it feel empty. As we walked the long aisles, we'd occasionally see a guy sitting cross-legged on the floor, his open laptop plugged into one of the giant machines—or perhaps sitting in a broken task chair, its backrest askew. It was uncomfortably noisy, the air was cold and dry, the perpetual half-light disorienting. And if a guy is on the floor in the first place, it's likely because something has gone wrong—a route is "flapping," a networking card has "fried," or some other mishap has befallen his network. He is mentally wrestling with the complicated equipment and physically uncomfortable. As we passed one tired-looking guy sitting in a small pool of light, like a troll, Troyer shook his head in sympathy. "Another poor schlep sitting on the floor." He yelled down the aisle: "I feel your pain!"

We passed the narrow room that stores the batteries that instantly provide backup power if the utility lines should fail. They were stacked up to the ceiling on both sides like drawers in a morgue. And we passed the generator rooms that take over from the batteries within seconds. Inside were six enormous yellow dynamos, each the size of a short school bus, each capable of generating two megawatts of power (creating the ten megawatts this one building needs at full capacity, with two extra for good luck). And we passed the 600-ton chillers used to keep the place cool: a giant steel insect of looping pipes each the diameter of a large pizza. For all the high-tech machines and uncountable quantities of bits, Equinix's first priority is to keep the power flowing and the temperature down: it's the company's machines that keep the other machines going.

Most of what I'd seen so far could have been in a data center anywhere. The equipment had arrived in crates with wooden skids, in cardboard boxes with CISCO stamped on the side, or on the back of a massive tractor-trailer, WIDE LOAD emblazoned on its bumper. But finally we came to a room where that wasn't true, and which I was excited to see above all else. Inside, the expanse of the planet—and the particularity of this place on it—was more explicit. The plastic name plaque on the door said FIBER VAULT 1. Morgan unlocked it with a key (no hand scanner here) and flipped on the lights. The small space was quiet and hot. It had white walls and linoleum floors, marked with a few scratches of Virginia clay. In the middle of the room was a wide-gapped steel frame, like three ladders fitted together side by side. Heavy-duty plastic tubes stuck out of the floor and rose to waist height, a half dozen on both sides of the rack, open on the top, each wide enough to fit your whole arm in. Some of these tubes were empty. Others erupted with a thick black cable, perhaps a fifth as wide as the tube. Each cable was marked with the telecommunications carrier that owned it, or used to, before it was bought or went bust: Verizon, MFN, Centurylink. The cables were attached to the wide frame in neat coils, and then looped up to the ceiling where each reached the highest level of ladder racking—the carrier innerduct. This was where the Internet met the earth.

There are different kinds of connection. There are the connections between people, the million kinds of love. There are the connections between computers, expressed in algorithms and protocols. But this was the Internet's connection to the earth, the seam between the global brain and the geologic crust. What thrilled me about this room was how legible it made that idea. We are always somewhere on the planet, but we rarely feel that location in a profound way. That's why we climb mountains or walk across bridges: for the temporary surety of being at a specific place on the map. But this place happened to be hidden. You could hardly capture it in a photograph, unless you like pictures of closets. Yet among the landscapes of the Internet, it was the confluence of mighty rivers, the entrance to a grand harbor. But there was no lighthouse or marker. It was all underground, still and dark—although made of light.

Troyer had long been sympathetic to my strange quest. He saw my excitement at coming to this small room that seemed to anchor the whole building—and with it much of the Internet—into the soft planet. "The whole idea of this building is that data can go in, and data can go out," he soliloquized. "It's the meet-me point where the Internet physically comes together to connect, so that it can become seamless and transparent to the end user. Where you're standing here, this happens to be the largest concentration of providers on one single campus in the US." Among the places where Internet networks connect, this was among the biggest—the nexus of nexus. Hot and still. I could smell it: it smelled like dirt.

I cracked a wide smile at the thought—at what a singular piece of Internet this was. And then Troyer knocked me down. As the sign on the door said, this was Fiber Vault 1. On the other side of the building was Fiber Vault 2. There were all the other buildings like it on the campus, each with its own multiple fiber vaults. This was the place, but so was that. And that. And that. The Internet was here, there, and everywhere.

We walked the length of the building back to the red silo, into the man-trap and out into the lobby. I slid my visitor's badge through the slot to the guard behind his bulletproof windows. We leaned on the single door that didn't look like anything and stepped out of the cool, dark building and into the hot, bright Virginia day.

Troyer raised his hand to the sky and moaned, "Ahh! Giant fiery orb!"

4

The Whole Internet

That evening in Washington, staying at my sister's house, I told my eight-year-old niece about what I'd seen at Equinix. She's an instant-messaging, YouTube-watching, video-chatting, iPad-swiping member of her generation: a digital native. And like most eight-year-olds, she's difficult to impress. "I saw the Internet!" I told her. "Or at least a very significant piece of it." I was accustomed to adults furrowing their brows at statements like this, skeptical that the Internet's physical reality could ever be so legible. But she didn't think it was strange at all. If you believe the Internet is magic, then it's hard to grasp its physical reality. But if, like her, you've never known a world without it, why shouldn't the Internet really be out there, something you could touch? It seemed to me that childlike wonder was a good way of looking at this world; it transformed everyday structures into monuments. And truly and fairly—even by rational adult standards—Equinix Ashburn was *the Internet* to a far greater degree than most anyplace else on earth could claim. The Internet is an expansive, near-infinite thing; but it's also astonishingly intimate. How reductive could I be—both in my imaginings of the Internet and my experience of its physical reality? What were the limits of precision?

Throughout this book I have been capitalizing "Internet," treating it as a proper noun. This increasingly goes against convention. If in the early days of the Internet, it was universally recognized as a unique thing and therefore deserving of its big letter, over time that novelty has been lost. As the website Wired .com explained when it switched away from the big "I," way back in 2004: "In the case of internet, web and net, a change in our house style was necessary to put into perspective what the internet is: another medium for delivering and receiving information."

But that's not how I see it. Not exclusively, at least. Because as soon as I began to engage with the Internet's physical presence— its places—it came into focus as a singular thing, however unusual and amorphous. By holding on to the proper noun, I'm also holding on to the idea that the Internet is demonstrably *there,* once you know where to look. And I don't mean only hidden away behind locked doors in unmarked buildings, but everywhere—in the wires circling the block and the towers of the skyline. This isn't to say I'm unaware of this idea's limits, and unwilling to acknowledge the way the networks slip from sight. Seeing the Internet like this demands a certain amount of imagination (occasionally crossing the line to hallucination). The writer Christine Smallwood is on to something when she points out that "the history of the Internet is a history of metaphors about the Internet, all stumbling around this dilemma: How do we talk to each other about an invisible god?" She weighs the relative merits of describing the Internet as a Tootsie Roll, a hot tub, a highway, or a plane, before finally acknowledging how ugly the Internet—the real Internet, the one I've been visiting—actually is. "I wish," she concludes, "the Internet looked like Matt Damon, or like lines of light written by an invisible hand in the night sky." And so she finds our old friend again: the amorphous blob, the infinite universe, vast, uncontainable, and expanding. The poets' metaphors all nestle together under the same starry night.

But comedians, I've noticed, tend to go in the other direction, toward "the Internet" as a single machine. On the episode of the television cartoon *South Park* called "Over Logging," the squat, obnoxious little characters faced a particularly extreme case of a familiar dilemma: the Internet breaks, everywhere. First they try to figure out if this is really happening, but "there's no Internet to find out there's no Internet!" a character deadpans. Soon enough "the Internet" itself appears on-screen, in the form of a machine the size of a house, looking suspiciously like a giant version of a home Linksys router, the blue one with the black front and little rabbit ear antennae in the back, lit by klieg lights in its underground bunker. Government agents in dark sunglasses are trying their best to fix it—at one point playing John Williams's famous five-tone motif from *Close Encounters of the Third Kind* at it, like a benediction. Eventually, one of the boys finds the solution: he climbs the aircraft-carrier-like steel ramp that leads up to the giant machine, goes around the back of it, unplugs it from an enormous socket, and plugs it back in. Salvation! The "flashing yellow light is steady green now!" a little guy cheers. Peace prevails.

The British sitcom *The IT Crowd* took the same joke to the opposite size extreme: the Internet wasn't a single big machine but a tiny little one. As an office prank, two of the IT guys convince a gullible colleague that "the Internet" is inside a black steel box hardly the size of a shoe, with a single red LED. It normally lives at the top of Big Ben—"that's where you get the best reception"—but with the permission of "the Elders of the Internet" they've been able to borrow it for the day, so she can use it for an office presentation. "This is the Internet?" she asks, incredulously. "The *whole* Internet? Is it heavy?" Her colleagues guffaw at her. "That's a bit of a silly question. The Internet doesn't *weigh* anything."

Watching these clips on YouTube, I felt a shudder of embarrassment. I seemed to be on a wild goose chase, looking for a world that few believed existed. But if only they really knew! The Internet isn't a little steel box, of course—not in toto, at least. But that isn't to say that there weren't a few steel boxes of vast importance (and that occasionally might need to be unplugged and plugged back in again). Sometimes the center of the Internet—or, at least, *a* center—is even more particular than a single building. So where were the biggest and most important boxes? And who were these "Elders of the Internet" calling the shots, assuming they existed at all?

The answer to both questions lay within the intimate world of "Internet exchanges." The terminology can be confusing, but for the most part the place where Internet networks meet is known as an Internet exchange, often abbreviated as "IX." The PAIX has this in its name; so too does Equinix, if more subtly. However, where things begin to get more slippery is that an "IX" might refer to either the brick-and-mortar building where networks connect to one another, or the institutions that facilitate that connection, whose equipment is often spread out among multiple buildings in a city. The important distinction is that an Internet exchange need not be a piece of real estate; it could be an organization. But there is still a physical thing—often with a single machine at its heart.

The rationale for an Internet exchange is straightforward, and not very different from the founding principle of MAE-East: get your packets to their destination as directly and cheaply as possible, by increasing the number of possible paths. If Ashburn serves this purpose on a global scale, there's also a need for smaller regional hubs. As the Internet itself has grown, that need has increased dramatically. Many engineers use the airport analogy: in addition to the handful of global megahubs, there are hundreds of regional hubs, which exist to capture and redistribute as much of the traffic within their area as is practical. But as with the airlines, the smaller nodes of this hub-and-spoke system are always pressured by the tendency toward global consolidation. As Internet networks (or airlines) merge, the big hubs become even bigger—sometimes with a significant loss of efficiency.

In Minnesota, the local network engineers refer to this as the "Chicago problem." Two small competing Internet service providers in rural Minnesota might find themselves sending and receiving all their data to and from Chicago, by buying capacity on the paths of one of the big nationwide backbones, like Level 3 or Verizon. But—as with a hub airport—the path of least resistance doesn't always make a lot of sense. In the simplest example, an email from the first network to the second network across town would travel to Chicago and back. Visiting the University of Minnesota's website, while sitting in Minneapolis, would entail a digital trip across state lines. But if you had a local Internet exchange, you could connect the two (or more) networks directly, often for only the cost of the equipment. The thing is, that might not even be worth the effort given the low cost of getting traffic to Chicago, and the low volume of traffic between those two particular networks. Sometimes it is easier to fly through Atlanta. But if that local traffic were to increase—and it always does—there's a point at which the elegance of interconnecting all of them, literally cutting Chicago out of the loop, is unmistakable.

In places that hang on to the Internet by a narrower thread, that threshold is more easily crossed, and keeping the traffic local is essential. Until recently, for example, the landlocked central African country of Rwanda depended entirely on satellite Internet connections, which were expensive and slow. If the few local ISPs weren't careful, an email heading across the capital of Kigali might end up having to make two 45,000-mile round-trips to space. In 2004, the RwandaIX opened to solve the problem, speeding up access to the local pieces of the Internet, and saving the expensive international bandwidth for traffic that was, in fact, international. And indeed it was this same idea that inspired the creation of what are now the biggest IXs globally.

In the mid-1990s, Internet networks everywhere didn't have

a "Chicago problem"; they had a "Tysons Corner problem." The traffic all went through MAE-East. The dozens of Internet exchanges now distributed around the world serve exactly this purpose. They range from behemoths like JPNAP in Tokyo, which posts astonishingly high traffic numbers but primarily serves intra-Japan communication; to the decidedly smaller Yellowstone Regional Internet Exchange, YRIX, which links together seven networks in Montana and Wyoming (curing them of their "Denver problem"). There's the MIX in Milan, the SIX in Seattle, the TORIX in Toronto, MadIX in Madison, Wisconsin, and—the solution to Minnesota's Chicago problem—the MICE, the Midwest Internet Cooperative Exchange. The vast majority of exchanges exist out of sight, often run as cooperative side projects for the "good of the Internet," and, despite their efforts at outreach, are known and appreciated only by the handful of network engineers who craft the routes across them.

But the largest Internet exchanges are different beasts altogether. Their participants aren't public-spirited groups of network engineers, but the Internet's biggest players globally. They are large, professional operations, with marketing departments and teams of engineers. The router manufacturers curry their favor, like sneaker companies courting the best athletes. And they are intensely competitive with one another, jockeying for the title of "world's biggest"—often by finding new ways of measuring. The two most used criteria are the amount of traffic passing through the exchange (both the peak at a given instant, or on average), and the number of networks that connect across it. In the United States, exchanges tend to be smaller; mainly because Equinix has been so successful in allowing networks to connect directly to each other. The big IXs, in contrast, rely on a centralized machine, or "switching fabric." The three biggest are all European: the Deutscher Commercial Internet Exchange, or DE-CIX, in Frankfurt; the Amsterdam Internet Exchange, or AMS-IX; and the London Internet Exchange, or LINX. Each has a live traffic graph on its website, along with a running tally of member networks. These three are an order of magnitude larger than the next tier—with the exception of the Moscow Internet Exchange, which has been breaking away from the pack. Watching their traffic statistics daily has the feel of a horse race, with one pulling away for a few weeks before another catches up, to the cheers of an invisible crowd. For months I tracked them closely, looking for changes and trends. And I quizzed network engineers and industry observers about which of the exchanges was qualitatively most important. "Well, Frankfurt's just huge," Alan Mauldin, Tele-Geography's analyst, had gushed to me about DE-CIX. "People have so much bandwidth going into bandwidth that it's just amazing." But Amsterdam was a close second and had been bigger for longer. And London, while posting lower numbers, trumpeted its "private" links, which moved much of its traffic off the exchange itself and onto direct connections, as in Ashburn.

But regardless of which was the single biggest, the idea of these big exchange points transfixed me. When I first set out in search of the Internet, I expected to find a loose arrangement of little pieces; it was all supposed to be distributed, amorphous, nearly invisible. I hadn't actually expected anything as grand and specific as a single booming box at the Internet's "center." That sounded more like science fiction. Or satire. But that's exactly what these big Internet exchanges were—except uncelebrated, under the radar, and somewhat oddly arranged, seemingly by-passing some world capitals while colonizing others. Their geography was peculiar: Why was Frankfurt big but not Paris? Tokyo but not Beijing? Did Germans spend more time online than the French? Or was it that the cities adhere to more fixed geographical patterns? Mauldin pointed out that Spain isn't a hub, and never will be. "It's a peninsula," he said. Geography was destiny, even on the Internet. Especially on the Internet.

But aside from looking at them like an analyst, measuring their size and commercial significance, I was curious about their physical reality. If geography mattered as much as it seemed to, that implied that these places worked on a smaller, more specific scale—a building or a box. To acknowledge that at all was to pull the Internet fully into the physical world. And once it was there I wanted to see it, touch it, ponder its corporeal presence. What were these big Internet exchanges *like*? A little black steel box with a single blinking light? A giant insectlike construction beneath klieg lights and behind barbed wire? The way Mauldin talked about DE-CIX in Frankfurt, it seemed like its "core" would be a sight to see—the Internet tourist's equivalent of the Grand Canyon or Niagara Falls or something else really big, certainly something worth crossing the ocean for. It was the Internet's busiest box. Surely that made it worthy of scrutiny, if not reverie. What might it say to me?

But there was also the security issue. These big exchanges made me nervous. Wasn't it dangerous for things to be concentrated like that? Or more to the point, was I being dangerous in seeking them out—with the intention of telling the tale? Certainly the existence of choke points like these contradicted the conventional wisdom about the Internet's redundancy. And the more I thought about this, the more paranoid I became.

Then, not long after my visit to Ashburn, arriving back in New York by plane, I had what you might call a brush with the law. As we taxied into the gate, the pilot came on the intercom and instructed us all, without further explanation, to stay in our seats. Two NYPD detectives, straight out of *Law & Order* central casting in loose-fitting suits, marched up the aisle, trailed by a uniformed patrolman. Everyone shot up to attention, pointing their chins over the tops of the seat backs.

At that moment, I thought they might be coming for me. The day before I had toured a particularly sensitive piece of Internet infrastructure, a building in downtown Miami known as the "NAP of the Americas." It serves as the Ashburn for Latin America, but it is also acknowledged to be a key interconnection place for military communications—a point that was hinted at on my tour. I had visited the building with permission and transparency about my project, but that didn't fully assuage my paranoia. The map of the Internet in my head had filled out. Was it possible, was there even *any* chance, that I now knew too much, without even knowing it?

With each row the cops passed toward the back of the plane the odds increased that they were coming for me. My wife and baby were with me, and the drama quickly played out in my head, the clichés cribbed from a hundred television shows: the handcuffs, the shouts ("I love you, call my lawyer!"), the outstretched arms. The cut to commercial.

But it wasn't me they wanted. It was the guy two rows back in a Mets cap, gray sweatshirt, and fifth-grader's backpack. He was silent as they led him off the plane. But he didn't look especially surprised. He did look like he'd had better days.

A little later I was interviewing a pair of veteran builders of the Internet's physical infrastructure at the offices of the private equity firm bankrolling their newest project. It was a lush place high above Manhattan, with thick carpeting and Impressionist paintings on the wood-paneled walls, and they were readily answering my questions about the important pieces of the Internet, and how their new piece would fit in among them. But I must have pushed a little too far.

The senior partner, in a double-breasted suit with a silk handkerchief in the pocket, interrupted his more loquacious colleague and glared at me across the oak conference table. "Let me ask you this," he said. "Are we creating through this book a road map for terrorists? By identifying the 'monuments,' as you refer to them, if they are known and damaged and destroyed, it's not just one building that goes down, it's the entire country that goes down, and is that a wise thing to be broadcasting to the world?" It was a startling accusation. Could my search for the physical infrastructure of the Internet be dangerous? Might the Internet be hidden for a reason?

I stammered a bit: I wasn't trying to hurt the Internet! I love the Internet! I tried to explain about my journalistic imperative, that only by making people more aware of these places will there be the wherewithal to properly protect them. I believed that, yet it didn't strike me as an argument that served his interests well. I turned the question back to him. I knew they wanted the attention; it would be good for business. So which was more important: that, or being quiet, perhaps quieter than his competitors?

He shrugged, before delivering a parting shot: "Do you want to be the guy who says, 'Here's what you attack to take down the country'?" And then he talked for another hour.

The truth was, his question stuck. In the course of visiting the Internet, when I arrived someplace new I often felt a kind of low dread that my journey was too eccentric to be palatable, that anyone I encountered would suspect ulterior motives, that I was a subject of suspicion. I was off the path, nosing around places few, if anyone, ever did. I wasn't so paranoid that I really believed I was being followed, but I didn't feel entirely comfortable either. After all, what was I really doing? ("Oh, you know, just sharing details about your local critical infrastructure with the world.")

But I shouldn't have worried. Inevitably, when I arrived at some unmarked building crucial to the functioning of the Internet, the same thing always happened. The veil of secrecy didn't descend, but lifted. Instead of stumbling around in the dark looking for the network, it often felt as if the lights had come up, and the more a person knew about the physical infrastructure of the Internet, the less concerned he or she appeared to be about its security. The "secret" locations I was interested in were not so secret after all. Whoever happened to be in charge happily led me around, and nearly always spent extra time to make sure I understood what I was looking at.

Over time I recognized that their openness wasn't merely polite, but philosophical—an attitude in part derived from the Internet's legendary robustness. Well-designed networks have redundancies built in; in the event of a failure at a single point, traffic would quickly route around it, so an engineer doing his job properly shouldn't be worried. More often, the biggest threat to the Internet is an errant construction backhoe or, in one recent well-publicized case, a seventy-five-year-old grandmother in the country of Georgia slicing through a buried fiber-optic cable with a shovel, knocking Armenia offline for twelve hours.

Yet above and beyond those practical concerns (or lack thereof) was a more philosophical rationale: the Internet is profoundly *public*. It has to be. If it were hidden, how would all the networks know where to connect? Equinix in Ashburn, for example, is unequivocally one of the most important network hubs in the world—as Equinix would be the first to tell you. (And if you enter "Equinix, Ashburn" into Google Maps, a friendly red flag will land square in the middle of the campus.) With the exception of certain totalitarian countries, a network doesn't have to apply to any central authority to connect to another network; it just has to convince that network it's worth its while. Or, even easier, just pay the network. The Internet has the character of a bazaar, with hundreds of independent players circulating around one another, working things out among themselves. This dynamic is at work physically, in buildings like the PAIX, Ashburn, and others. It's at work geographically, as networks move to complement one another's regional strengths. And it's at work socially, when network engineers break bread and drink beer.

When we're sitting in front of our screens, the path by which everything comes to us is entirely obscured. We might notice that one page loads faster than another, or that a movie streaming from one site always looks better than one from another—a result, very likely, of fewer hops between the source and us. Sometimes this is obvious; I recall planning a trip to Japan, and waiting as local travel pages loaded like molasses. Other times it takes an extra leap of understanding; video-chatting with a friend in another city, I couldn't get over how good the quality was until I remembered that she had the same home Internet service provider. The stream never had to leave the network. But by and large, when we enter an address in our browser, or an email arrives in our inbox, or an instant message flashes on the screen, there's no clue whatsoever as to the path it took to get there, how far it traveled or how long it took. From out here, the Internet appears to have no texture, no grain; with rare exceptions, there's no "weather"—conditions don't change day to day.

Yet looked at from within, the Internet is handmade, one link at a time. And it's always expanding. The constant growth of Internet traffic requires the constant growth of the Internet itself, both in the thickness of its pipes and the geographic reach of individual networks. For the engineers, that means a network not busy being born is busy dying. As Eric Troyer said about Ashburn, "The goal of coming into sites like ours is to create as many vectors out to the logical Internet as you can. The more vectors, the more reliable your network becomes—and generally the cheaper it becomes because you have more ways to send your traffic."

———

So the Internet is public *because* it's handmade. New links don't just happen according to some automated algorithm, they need to be created: negotiated by two network engineers, then activated along a distinct physical path. That's hard to make happen in secret.

Making those connections between networks is known as

"peering." In the simplest terms, peering is the agreement to interconnect two networks—but that's like saying "politics" is merely the activity of government. Peering implies that the two networks involved are "peers," in the sense that they are of the same size and status, and therefore exchange data on more or less equal terms, and without money changing hands. But figuring out who's your peer is a touchy business in any context. Inside the Internet, it's made more complicated when peering can also mean "paid peering"—when something with a clearer value than data is added to tip the scales in one direction or another. In its subtleties and nuances, peering has a Talmudic quality, with a body of laws and precedents that are ostensibly public but require years of study to be properly understood. The consequences are huge. Peering allows information to flow freely across the Internet—by which I mean both liberally and at low cost. Without peering, online videos would clog the Internet's pipes—YouTube might no longer be free. And service providers would accept less reliability in the name of lower costs. The Internet would be more brittle and expensive. Given those stakes, nowhere is the process of Internetworking more intense, and more fraught with occasional drama, than among the network engineers loosely known as "the peering community."

I went to observe them firsthand at one of the thrice-annual meetings of the North American Network Operators' Group, or NANOG, at the Hilton in Austin, Texas. When I arrived, the hotel lobby was filled with men in jeans and fleece, chatting quietly with one another across the tops of laptops festooned with bumper stickers. These are the wizards behind the Internet's curtains—although plumbers might be just as good an analogy.

What they do certainly seems like magic. Collectively, they command a global nerve system of astonishing capabilities, even if most of the time its daily operations are mundane. But mundane or not, there's no doubt we're fearsomely dependent on the body of highly specialized knowledge that only they possess. When things go wrong in the middle of the night on the Internet's biggest pipes, only the NANOGers know how to fix it. (And it's a stale joke at the conference that if a bomb went off in its midst, who would be left to run the Internet?) They aren't primarily bureaucrats or salespeople, policymakers or inventors. They are operators, keeping the traffic flowing on behalf of their corporate bosses. And on behalf of one another. The defining characteristic of the Internet is that no network is an island. Even the most crack engineer is useless without the engineer who runs the next network over. Accordingly, people don't come to NANOG for the formal presentations. They come for the networking opportunities—and not "networking" as a figure of speech. Plenty of business cards were exchanged at the conference I attended, but so were Internet routes. A NANOG meeting is the human manifestation of the Internet's logical links. It exists to cement the social bonds that underscore the Internet's technical bonds—a chemical process aided by ample bandwidth and beer.

The typical NANOGer will have the job title of "engineer" preceded by one of a handful of qualifiers like "data," "traffic," "network," "Internet," or, occasionally, "sales." He—and nine out of ten attendees in Austin were men—might run the Internetwork of one of the biggest and most familiar suppliers of Internet content, like Google, Yahoo!, Netflix, Microsoft, or Facebook; one of the biggest owners of the Internet's physical networks, like Comcast, Verizon, AT&T, Level 3, or Tata; or one of the companies variously serving the Internet's inner workings, from equipment makers like Cisco or Brocade to cell-phone manufacturers like Research in Motion, to volunteer delegates from ARIN, the Internet's contentious, United Nations–like governing body. Jay Adelson was a NANOG fixture until he left Equinix, and Eric Troyer rarely missed a meeting. Steve Feldman—the guy who built MAE-East—was the chair of the NANOG steering committee.

If for most of us a given bit's journey across the Internet is opaque and instantaneous, for a NANOGer it is as familiar as a walk to the grocery store. At least in his own Internet neighborhood he will know each link along the way. He can invariably diagram the logical links and, in all likelihood, picture the physical ones. He may have set it up himself, configuring the routers (perhaps even unpacking them from their original boxes), ordering the appropriate long-distance circuits (if not showing where they should be dug into the ground), and continually fine-tuning the flows of traffic. Martin Levy, an "Internet technologist" at Hurricane Electric, which runs a good-sized international backbone network, keeps a photo album of routers on his laptop, alongside pictures of his son. These are the people with the best mental maps of the Internet, the ones who have internalized its structure beyond all others. And they're also the ones who know that its proper functioning—that every move you make online—depends on a clear and open path across the whole Internet, from end to end.

The peering people divided into two camps: those looking for new networks to connect to their network; and the facility owners and Internet exchange operators who compete to host those physical connections in their buildings. The highest powered of both sets tended to be more extroverted, bopping around during the coffee breaks slapping hands and handing out business cards. They were better dressed, and they bragged about how they could hold their alcohol. Take, for example, the peering link between Google and Comcast, the big US cable company. The YouTube videos, Gmail emails, and Google searches of Comcast's fourteen million customers would, as much as possible, use the direct link between the companies' networks, avoiding any third-party "transit" provider. Physically, that Comcast-Google link would be repeated a handful of times in places like Ashburn and the PAIX (and indeed in those two specifically). But socially, it is visible in the relationship between the peering coordinators, Ren Provo at Comcast and Sylvie LaPerrière at Google—two of the few women at the conference. Provo, whose official title is "Principal Analyst Interconnect Relations," worked the crowd at NANOG in her Comcast bowling shirt, asking about people's families and yelling jokes across the room. Her husband, Joe, is also a network engineer and high up in NANOG's volunteer bureaucracy, making the Provos the conference's unofficial power couple; many NANOGers speak fondly of their wedding weekend. LaPerrière is a charming French Canadian whose business card reads "Programme Manager." She seems to be universally adored, albeit tempered by an undercurrent of fear at her power. If you run a network, you want good links to Google—if only so your customers don't complain their YouTube videos are jittery. LaPerrière makes it easy for them. For the most part her job is to say yes to all comers, since those good links are in Google's interest as well. Their peering policy is "open"; in the macho jargon of NANOG that makes Google a "peering slut" (a term LaPerrière loathes). Not surprisingly, LaPerrière and Provo are good friends, and I often spotted them huddling in the hotel hallways. Their relationship helps smooth the way across a technically complex and financially wrought minefield of variables.

"You can tell if your friend is telling the truth or not," LaPerrière explained to me, before quickly adding that friendship has its limits. "In the long run you'll only be successful if you're truly representing your company, and you make it very clear that this is your company's policy, not the 'Sylvie policy,'" she said. "Friendship in my book doesn't play a role at all. It just makes the interactions nicer." Yet her protestations only make the broader point: a connection between networks is a relationship.

And peering can get nasty. Occasionally a major network will "de-peer"—literally pull the plug on a connection and refuse to carry its combatant's traffic, usually after failing to convince the other network that it should be paying them. In one famous de-peering episode in 2008, Sprint stopped peering with Cogent for three days. As a result, 3.3 percent of global Internet addresses "partitioned," meaning they were cut off from the rest of the Internet, according to an analysis by Renesys, a company that tracks Internet traffic flows and the politics and economics of connection. Any network that was "single-homed" behind Sprint or Cogent—meaning they relied on the network exclusively to get to the rest of the Internet—was unable to reach any network that was "single-homed" behind the other. Among the better-known "captives" behind Sprint were the US Department of Justice, the Commonwealth of Massachusetts, and Northrop Grumman; behind Cogent were NASA, ING Canada, and the New York court system. Emails between the two camps couldn't be delivered. Their websites appeared to be unavailable, the connection unable to be established. The web had broken into pieces.

For Renesys, which makes a business out of measuring the quantity of Internet addresses handled by each network and reading the tea leaves as to their quality, a "de-peering event" like that is an amazing moment, like the lights being flipped on at a club. The relationships are revealed. The topography of the Internet is inherently public, or else it wouldn't work—how would the bits know where to go? But the financial terms that underpin each individual connection are obscured—just as an office's physical address is public while the details of its lease are private. The lesson Renesys was selling from this analysis was that anyone serious about their Internet should be "intelligently multi-homed." Meaning: don't roll all your eggs through one network. The network engineers' credo is "Don't break the Internet." But as Renesys's Jim Cowie explained, that cooperation goes only so far. "When it gets to a level of seriousness, people get very quiet. There's a huge amount of money and legal exposure at stake."

Traditionally, peering has been dominated by an exclusive club made up of the biggest Internet backbones, often known as the "Tier-1" carriers. In the strictest definition, Tier-1 networks don't pay any other network for a connection; others pay them. A Tier-1 network has customers and peers, but it doesn't have "providers." What results is a tightly interconnected clique of giants, often whispered about as a "cabal." Renesys tracks the relationships among them by "reading the shadows on the wall," as Cowie put it, created by the routes each network broadcasts to the Internet routing table—the signs that say "this way to that website!" But because the exact agreements between networks are private even if the routes are public, the precise list of Tier-1 providers can be hard to write. In 2010, Renesys identified thirteen companies at the top of the heap, and four at the very top: Level 3, Global Crossing, Sprint, and NTT. But in 2011, Level 3 purchased Global Crossing, in a deal valued at $3 billion—so then there were three.

However, peering has been evolving in recent years. As the Internet has grown, the practice has become increasingly distributed. It's become more cost-effective for smaller networks to peer among themselves, in part because many smaller networks have become pretty big. And while peering used to be more common among regional networks (like those guys in Minnesota), it's now more frequently seen at a global scale. These new peering players are different in that they are not primarily "carriers," meaning they're not in the business of carrying other people's traffic; instead, they're plenty worried about their own traffic. It's like the Internet version of a university or company operating a shuttle bus between campuses, rather than relying on public transportation—or the huge companies that will do the same with a private airplane between cities. When there's enough traffic between two points, it becomes worth it to move it yourself.

They include some of the Internet's most familiar names, including Facebook and Google. In recent years, both have put enormous resources into building out their global networks, in general not by laying new fiber-optic cables (although Google did partner on the construction of a new cable under the Pacific) but by leasing significant amounts of bandwidth within existing cables or buying individual fibers outright. In that sense, a network like Google's or Facebook's will be logically independent on a global scale: they each have their own private pathways, traveling within the existing physical pipes. The crucial advantage of this is that they can store their data anywhere they choose—primarily in Oregon and North Carolina, in Facebook's case—and use their own networks to move it around freely on these private pathways parallel to the public Internet.

Their networks are headed directly to Internet exchanges (not your home)—an architecture that only makes the exchanges more important. If you're going to bother building out your own network, it needs worthwhile places to go: good regional distribution nodes. A network will connect directly into a place like Ashburn, where its owner will hang a shingle to announce its willingness to interconnect with other networks. In some cases, it's actually a shingle: a printed placard looped on to the outside of a cage, to attract the attention of visiting network engineers. More often, it's a virtual shingle: a listing on a website called PeeringDB, or just an information page of its own, as Facebook has.

Facebook.com/peering isn't behind any password, or within some proprietary database. It's wide open—as exposed as your cousin's vacation photos. A brief paragraph at the top describes Facebook (for the benefit, one supposes, of any peering coordinators from Mars): "Facebook is a social utility that helps people communicate more efficiently with their friends, family and coworkers." Then it lays out the company's peering m.o.: "We have an open peering policy welcoming the opportunity to engage in peering with any responsible BGP speaker in an effort to improve the experience of our millions of users throughout the globe." Being a "responsible BGP speaker" means you know how to configure a big Internet router and can be counted on to fix it fast if you screw up. Its "open peering policy" makes Facebook a classic peering slut, happy to connect with all comers. Then there is a table that shows *where* you can connect, listing sixteen cities around the world, the particular Internet exchange in those cities, the IP address (like the Internet's postal code), and the capacity of the "port" at that location.

The first time I saw Facebook's list—during a coffee break at NANOG—my eyes widened. For months I'd been talking to network engineers and facility owners, polling them about the Internet's most important places, working up a judgment about where, exactly, one looks for the Internet. And then here, wide open to the world, was exactly that—at least according to Facebook (the world's second-most-visited website, after Google).

Facebook is not in the business of delivering Facebook pages to people's homes, offices, or cell phones; it relies on other networks to do that. This page said that if you run one of those networks, and you're "responsible," Facebook will connect to you, either directly (router plugged into router like at the PAIX or in Ashburn) or via a central switch (at an Internet exchange). It's Facebook's attitude that the more connections, the merrier, which makes this public information in the same way that American Airlines will happily tell you where it flies. So if you're a small ISP in, say, Harrisburg, Pennsylvania, and you've noticed that a meaningful percentage of your traffic is coming from Facebook, and you've built your network back to Ashburn anyway, you'd strongly consider asking if you can plug in directly.

Facebook's peering coordinator will probably say yes. And your customers will notice that Facebook loads faster than almost anything else.

Yet what's most telling about Facebook's peering list is how short it is. The global capitals aren't surprising: New York, Los Angeles, Amsterdam (at AMS-IX), Frankfurt (at DE-CIX), London (at LINX), Hong Kong, and Singapore. The big US cities—Chicago, Dallas, Miami, and San Jose—are also to be expected from an American company. But the small US cities put in stark relief the unique geography of the Internet. When else is Ashburn, Virginia, on the same list as London and Hong Kong? Or Palo Alto? Vienna, Virginia (also on the list), is next door to Tysons Corner, which (for the moment) still has enough gravity to draw a crowd. It's clear that the geography of these buildings can partly be understood on a global scale: the Internet follows its users, meaning all of us, to where we live. But it's also clear that if you zoom in on the map, those broader forces fall away and are instead replaced by the ad hoc decisions of a small number of network engineers, each searching for the most technically and economically efficient places to connect. Palo Alto or San Jose? Ashburn or Vienna, Virginia? The thing about peering is that its influence cuts both ways: networks want to be at places where there are lots of networks. Facebook's choices of locations are therefore both a response to the growth of a given location, and a seed to its future growth. Or maybe it just likes the blue lights at Equinix. Or the beer in Amsterdam. Or both.

The chatter about peering came to a head on the last full day of the NANOG meeting in Austin, at a session listed somewhat cryptically on the agenda as the "Peering Track." Its late timing was deliberate, when the network engineers were hopped up like schoolkids on a Friday afternoon. Inside the meeting room, the banquet chairs had been arranged in a circle, ostensibly to facilitate discussion, but more, it seemed, to create a gladiatorial atmosphere.

This was peering speed dating: Internet exchanges advertised their size and prowess, and peering coordinators advertised to one another. They called it "peering personals." A successful pitch in front of this crowd might lead to a chat afterward, and a new route after that. Perhaps you run a data center in Texas, but you happen to have a big Danish website as your customer. Peering with a Danish ISP might get a lot of traffic off your hands—enough to make it worth meeting that ISP in Ashburn. That's exactly what Nina Bargisen, a network engineer at TDC, the Danish telephone company, had in mind when she put forth this simple plea: "I have eyeballs eyeballs eyeballs," she said. "For all of you with content, please send me an email."

Dave McGaugh, a network engineer at Amazon, did his best to dispel his colleagues' expectations that his traffic ratios were skewed in his favor: "We are outbound heavy but not to the extent people might expect," he pleaded. Will Lawton, the representative from Facebook, was understandably aggressive in his offer; Facebook, after all, has a lot of news to share. Facebook's typical "ratio of egress to ingress" was 2:1, he said, assuring any skeptics that he'd be accepting plenty of their traffic (like uploaded photos) for every bit he'd be sending back (the viewed photos). Across the board, the message from the peering coordinators was "peer *with* me."

The message from the exchanges was "peer *at* me," make that physical link in my city—make my place your place. The competition was intense, especially among the biggest. The audience was hushed as in back-to-back presentations the London, Amsterdam, and Frankfurt exchanges each highlighted their recent growth, offered a few words about their robust infrastructure, and finished with a pitch about the importance of their place in both the physical and logical worlds. The smaller exchanges followed, left only with the comparative advantage of their real-world geography—their ability to solve each network's particular "Chicago problem." As Kris Foster, the representative from TORIX, in Toronto, offered, "If you've got fiber routes from New York to Chicago through Toronto, maybe you should think about stopping." It wasn't a bad suggestion. It might make your network more efficient, but at the least it was a way to be "multi-homed," a salve against fragmentation.

But for the most part, the gravitational pull of the biggest exchanges was powerful enough to overcome that geographic diversity. The network effect was unassailable: more networks disproportionately concentrated in fewer places. The result was a striking gap between the average-sized Internet exchanges and the behemoths. "Places like AMS-IX and LINX are a routing engineer's paradise," Renesys's Cowie explained. "They are full of hundreds of organizations that are begging you, 'Please look at my routes, study me!'" That's irresistible for a NANOGer. The big exchanges become bigger, and it seems likely they will stay that way, growing in proportion with one another. That made my job a little easier: the Internet's map was, for the moment, fixed. My itinerary was clear. If I wanted to see those singular boxes at the center of the Internet, then Frankfurt, Amsterdam, and

London were the places to go. What I didn't know yet was how different they would be. Or if they accepted visitors.

That evening Equinix sponsored a party at an Austin music club with a huge roof deck. A large "E" was projected on the dance floor and a woman at the door handed out guitar-shaped Equinix key chains, in honor of the city's famous music scene. It wasn't over-the-top extravagant, but it was the biggest party on the closing night, with enough free drinks to ensure that none of the NANOGers would miss it. For Equinix, hosting it was a no-brainer. The hundred or so network operators milling around drinking beer each represented a network, and they were more likely than not all Equinix customers—if not several times over. Even better, if two network engineers met at this party and then decided to peer, the decision would be consummated with a "cross-connect," a cable from one cage to another—which meant recurring revenue for Equinix. The key chains were the least it could do.

I had my own agenda. I didn't have a network (excepting the dusty one behind the couch), but rather an image of all the networks, a giant imaginary map I had been steadily filling in. During the peering session, the AMS-IX presentation had been given by a young German engineer with a shock of red hair. Her boss, a jovial and slightly rounded Dutchman named Job Witteman, had watched quietly from the side, looking a bit like the Godfather, leaning back in his chair. He was the guy I wanted to talk to. NANOG attracts a brainy crowd with strong opinions, not always gracefully expressed. The session Q&As are almost always contentious. Shouting matches aren't uncommon. (The self-consciously combative tone of the peering session was meant to neutralize some of that.) But Witteman gave the impression of rising above all that drama, an elder statesman with his head above the fray. He didn't seem like an engineer.

"I never touched a router in my life!" he yelled over the music, when I asked, at the party. "I know how to switch on a computer, that's about it. I know what a router can do and how it functions—but don't ask me to touch it. That's what other people are for." He ran one of the largest Internet exchanges, but he'd always avoided learning the nitty-gritty of networking. The strategy worked; it was one less thing to argue about—and AMS-IX was known for its technical competence. Witteman sketched its history for me. Like many exchanges, it was founded in the mid-1990s as an offshoot of its nation's early academic computer networks. But unlike most, it was quickly professionalized. Rather than letting volunteers handle technical support, AMS-IX treated the task from the beginning with the same precision and planning as the Dutch treat everything else. "The whole purpose of us building a business back then was to *become* professional," Witteman said. "We didn't want to be like, 'Who's in charge today?'"

From early on, he took an equally Dutch approach to creating a marketplace. As AMS-IX shook off the mantle of the computer science departments, its spirit became intensely commercial—yet communal. "It's somewhere in the genes in the Netherlands. We are good at trading organizations, at exchanges—like tulip bulbs and whatnot. We don't need to buy and sell, but we like to create the marketplace," Witteman said. A very open marketplace. On the AMS-IX, as on the streets of Amsterdam, you could do whatever you want—so long as it didn't bother anyone else. "We've always been the company where we say, 'This is your platform, you pay for it, your port size determines the amount of traffic you can flow over the exchange, but we don't look into it, we don't care, we don't mind.'" It struck me as a pure expression of the Internet's publicness, this interconnected group of autonomous networks left to their own devices in a carefully managed setting. It also recalled the Internet's founding ideas, vaguely Californian, of live and let live, "be conservative in what you send and liberal in what you accept." To a certain extent this was true not only of AMS-IX but all Internet exchanges.

It was an openness that came with risks. There's little doubt the scourges of the Internet—child pornography chief among them—cross the AMS-IX, and others. But Witteman was adamant that it wasn't his business to prevent it, no more than the post office was responsible for what it carries. When the Dutch police once asked to plug in tapping equipment, Witteman carefully explained that it wasn't possible—but they could become a member of the exchange if they wanted and peer individually with the ISPs they're responsible for policing. "Now they're paying for their port and everybody is happy," Witteman said, taking a swig and raising his eyebrows.

As we talked, a tall man with a trim beard and spiky hair joined us, thumbing at his smartphone purposefully while he waited for a break in the conversation. When it came, he turned the little screen toward Witteman and shook his head with feigned surprise. "Eight *hundred,*" he said. Witteman responded with wide eyes, as if with mixed emotions. The tall man was Frank Orlowski, Witteman's counterpart at DE-CIX in Frankfurt. He meant eight hundred gigabits per second, his exchange's peak traffic that afternoon, another new record. More than Amsterdam. Ten times more than Toronto.

Clearly the competition among these big exchanges was at least a little personal. Orlowski and Witteman were on the circuit together—sometimes co-conspirators, oftentimes friendly competitors. They'd both made the journey across the Atlantic to Texas, having crossed paths just a few weeks before at a similar event in Europe. Witteman was undoubtedly jealous of DE-CIX's growth, but he was equally proud and amazed that their baby—by which I mean the Internet itself and the exchanges that sit at its center—had grown so big. As I listened to them, I loved how intimate it made the Internet's infrastructure seem, how the queries and messages of an entire hemisphere could be understood in the clink of beer bottles in a bar in Texas. The social binds that tied the physical networks together were visible, not just among Witteman and Orlowski but across the whole crowd at the party. I can't say I was surprised that the Internet was run by wizards—it had to be run by *somebody*. But I was surprised by how few they were.

But what about the places, the physical components, the hard ground? Was the Internet as concentrated as it had seemed during the peering session, or looking around the bar that night? At NANOG, the engineers were focused on just a handful of cities, with a clear hierarchy. Was this the Internet's geography? Frankfurt and Amsterdam were as far apart as Boston and New York (which isn't very far); Witteman and Orlowski ran equally professional operations. They both put in the effort to rub shoulders here, thousands of miles from home, encouraging engineers to peer at their exchanges. What would make a network engineer choose one over the other? Would I be able to tell the difference between the two places?

These big exchanges seemed a distillation of the Internet's essence: single points of connection that sparked new connections, like a hurricane gaining power over the ocean. I wanted to be reassured that even though the Internet made places less relevant, its own places still mattered—and with it, perhaps, the whole corporeal world around me. I'd come to NANOG to meet the people who individually ran the networks and collectively ran the Internet. But what I really wanted was to see the places where they meet, to somehow get closer to understanding the physical geography of my virtual life. Witteman and Orlowski's pieces of the Internet were inherently rooted in their own places—as distinct as their own national identities. Where did all those wires lead? What was there actually to see?

I told Witteman and Orlowski about my journey, before raising what I hoped was the obvious question: Could I come see for myself? "We have no secrets," Witteman said, cartoonishly looking inside his jacket flaps to emphasize the point. "Anytime you want to come to Amsterdam."

Orlowski looked down at us both and nodded in agreement. I was welcome in Frankfurt as well. We toasted with our Equinix beers.

On the late winter day I arrived in Germany, the gray sky perfectly matched the steely bank towers that rose up beside the river Main. I spent a jet-lagged Sunday afternoon exploring Frankfurt's quiet center. In the cathedral, I saw the side chapel where the kings of the Holy Roman Empire would gather to choose an emperor. From that single room, the news was sent out across the land. Nearby was a more contemporary landmark: the large statue of a blue-and-yellow euro symbol, famous as the backdrop for news reports about the European Central Bank. Both chapel and euro hinted at the particular spirit of this place: Frankfurt has always been a market town and communication hub, a sternly self-important city.

That evening I had dinner at a hundred-year-old restaurant with a friend, an architect and Frankfurter. Over beef with traditional green sauce washed down with pilsner, I pressed him to help me understand the city, and how its big piece of Internet fit in with the whole. But he mostly demurred. Frankfurt is not a place given to romantic aphorisms. It is short on homegrown anthems and atmospheric poems. It does not often appear in films. Among its most famous sons are the family Rothschild, the great German-Jewish banking dynasty (whose success, appropriately enough, came from strategically fanning out around the world—and using carrier pigeons for speedy long-distance communication), and Goethe (but he hated the place). Among Frankfurt's greatest contributions to twentieth-century culture is the "Frankfurt kitchen," a kitchen design of supremely utilitarian character, even by the standards of the Bauhaus (it sits in the collection of the Museum of Modern Art). If anything, Frankfurt is best known as a place for trade fairs (a role it's played since the twelfth century), like the Buchmesse for books and the Automobil-Ausstellung for cars, and for its big airport, among Europe's largest hubs. So I wasn't entirely surprised when my friend eventually came to a simple and resonant observation: Frankfurt was a transient city, a place where people did their business and then left. Despite its five million inhabitants, Frankfurt was not a place to truly live. And that turned out to be true for the bits as well.

The next morning I went to DE-CIX's offices in a brand-new building of glass and black steel, overlooking the Main in a stylish neighborhood of design stores and media companies, nearer to the shipping docks than the banks. The DE-CIX network engineers worked in an open room in the back, their desks set among whiteboards on wheels, scattered purposefully like the trucks on an airport tarmac. But this was only DE-CIX's administrative home. The "core switch"—the beating heart of the exchange, the big black box through which those 800 gigabits per second of traffic flows—was a couple of miles down the road, and its backup was equidistant in the other direction. We'd pay our respects after lunch.

First, I sat down with Arnold Nipper, DE-CIX's founder, chief technology officer, and something of a father to the German Internet. He looked the part, dressed like an esteemed computer science professor in a work shirt and jeans, his smartphone and BMW keys set on the conference table in front of him. With twenty-five years of experience explaining computer networking to the rest of us, he spoke slowly and precisely, his English accented a bit like Sean Connery.

In 1989, Nipper established the first Internet connection for the University of Karlsruhe, a technology powerhouse, and later was a lead developer of Germany's national academic network. When the commercial Internet came into existence in the early 1990s, Nipper became chief technology officer of one of Germany's first ISPs, Xlink, where he faced a familiar headache: MAE-East. "Every packet had to go across really expensive international links to the NSFNET backbone," Nipper said, between sips of an espresso. In 1995, Xlink teamed up with two other early ISPs, EUnet in Dortmund (home of another prominent university computer science department) and MAZ in Hamburg, intending to take the transatlantic crossing out of the loop by linking their networks together on German soil.

The Deutscher Commercial Internet Exchange was founded with a ten-megabit interconnection—about 1/100,000th its current capacity—and a hub installed on the second floor of an old post office building near the center of Frankfurt. This first incarnation of DE-CIX was far from grand, but it changed everything. For the first time, Germany had its own Internet—its own network of networks. "With the invention of DE-CIX we had only to cross national links," Nipper said.

But why Frankfurt? Nipper admits the decision to put the hub here—a decision that today exerts a sunlike gravity on the geography of the whole Internet—was somewhat unconsidered. Mostly it was because the city was at the rough geographic center of those first three participants, a quality from which it's often benefited. And it helped that Frankfurt was the traditional center of German telecommunications and, of course, Europe's financial capital. Yet it would be wrong to conclude that in this case the Internet followed finance—or followed Deutsche Telecom. That first hub wasn't planned for growth, or at least not the rocket ship that was about to take off. The Internet's expansion was ad hoc, as always.

In the decade and a half since its founding, a lot has changed, not only at DE-CIX but across Europe—and it's Europe's changes that have primarily driven DE-CIX's growth. As Internet use in the former Soviet republics catches up with the West, DE-CIX has aggressively pitched ISPs there, offering connections to networks from around the world, for less money than London or Ashburn and with a richer mix of global carriers to boot, particularly from the Middle East and Asia. It's not that the former republics have so much to say to Frankfurt; it's that Frankfurt, in large part thanks to DE-CIX, is the easiest place for them to hear from the rest of the world. The long-distance infrastructure is here: the major European fiber routes converge along the Main, leveraging the same geographic centrality that makes Frankfurt's airport one of Europe's largest hubs, and has always made Frankfurt a hub. But mostly the networks are following the economic truth that it's cheaper and more reliable to "tether" yourself into a big exchange than rely on someone else to bring everything back for you. That truth is self-perpetuating.

For example, Qatar Telecom, based in the tiny Persian Gulf state, has established beachheads in a familiar list of places around the world: Ashburn, Palo Alto, Singapore, London, Frankfurt, and Amsterdam. Its network carries both voice and private data services for corporations, but when it came to public Internet traffic, undoubtedly the easiest thing would have been to buy "transit" at its doorstep in Qatar from one of the big international providers—perhaps Tata, the Indian conglomerate, which has robust links to the Persian Gulf. But that would have meant leaving to someone else the business of bringing the Internet there. Instead, Qatar Telecom has installed its own networking equipment in those major exchange points and leased its own private fiber-optic circuits back to the Gulf. No wonder the exchanges are competitive with one another. They're all lobbying for more peering—but really they're hoping that "more" really means "here."

In Frankfurt, there was still the matter of seeing what that "here" looked like. After lunch, Nipper drove us east along the river in his little station wagon, onto a narrow street in a densely packed neighborhood of sturdy old warehouses. DE-CIX's primary "core" is in a building operated by a colocation company named Interxion, a European competitor of Equinix's. It opened in 1998, with DE-CIX as one of its first tenants (and still the only exchange). In contrast with the suburban expanse of Virginia, its parking lot was narrow and tidy, with cobbled pavement and manicured bushes, surrounded by a cluster of low, white buildings studded with security cameras. We squeezed in beside a gunmetal-gray van, its rear doors swung open to reveal a rack of tools, "Fibernetworks" painted on its side. Nearby, a red-hard-hatted worker rode a jackhammer into the ground, and a pair of technicians were busy taking apart an automatic doorway.

"You can see business is healthy," Nipper said, nodding at the construction. I asked if he was a big customer. "We are an important customer," he corrected, coolly. They're the honey trap, and treated well because of it; many Internet exchanges often won't have to pay rent for their own equipment. We were waiting for a guard to escort us, and Nipper thought the security rigmarole was overkill. "It's only a telecommunications hub," he said. "Data going through here is transient, you see. It's not like a disaster recovery center for a bank, where data is stored—that really has to be secure. Even if this completely fails, it will of course cause an impact on the Internet, but perhaps no email will be lost, just your browser will hang for a second or so, then everything is re-routed." The DE-CIX core, for its part, is designed to "fail-over" to its backup location across town within ten milliseconds. Easy come, easy go.

When our security guard finally arrived, we raced to keep up with her as she crossed the parking lot. Nipper buzzed us into the bare lobby of the data center with his keycard and then swiped us again into a second anteroom, all white walls and fluorescent lights. There we came upon a bit of hubbub. A construction worker in blue coveralls was stuck inside the glass cylinder of the man-trap, like a squid in a test tube. The fingerprint scanner wasn't recognizing his dirty hands and had locked him in, to the jeers of his coworkers on either side of the glass. Compared to Equinix, it all felt less cyberrific, more Teutonic prison. When it was our turn, the guard barked into her radio, and both sides of the man-trap snapped wide open. A loud klaxon sounded, and she hustled us through the door.

Inside, the equipment cages weren't cages at all, but full enclosures of beige steel that reached up toward the high, open ceiling, lined with the familiar tracks of yellow fiber-optic cables. Each space was labeled with a numeric code divided by a decimal point—no signs or names anywhere. In Europe it's typical for power and cooling to run underneath the floor, whereas in the United States it often runs up above. When we came to DE-CIX's space, the guard handed Nipper a key on a green rubber bracelet, as if we were visiting a safe-deposit box at a bank. The room had the size and character of an airport bathroom, all tidy and beige. As in Ashburn, the roar of the machines was deafening, and Nipper shouted to be heard.

"One of our principles is to make everything as simple as possible, but not simpler," he said, quoting Einstein. DE-CIX's "typology," in engineering terms, is its own design. The connections from each of the nearly four hundred networks that exchange traffic here are aggregated together—"muxed up"—into a handful of fiber-optic cables. A "fiber protection device" then acts like a two-way valve, directing the flow of data between DE-CIX's two core switches, the active and the "hot standby," here and across town. The job of the core is to direct incoming traffic out the correct door toward its destination. The majority of the light travels through the active path to the live core, but 5 percent of the signal is refracted toward the backup, which is always up and running.

"All these boxes have been communicating. If one link fails, it tells all of the other boxes to switch over at the same time, and they do it within ten milliseconds," Nipper said. He tests the system four times a year, switching between the two cores during the quiet hours early on a Wednesday morning. Despite its international customers, traffic across DE-CIX plots a cresting wave, rising throughout the day and peaking midevening, German time, as everyone settles in at home with their web videos and shopping. As Nipper explained all this the guard watched us carefully from the end of the aisle, like a stalker in a grocery store.

Nipper saved the core itself—the crown jewels—for last, working the key on the green bracelet into the lock on the cabinet and then opening it with a playful flourish. I skipped a breath as I took it in: a black machine in a standard-sized rack; yellow fiber-optic cables sprouting from it like spaghetti from a pasta maker; dozens of busily blinking LEDs; a printed white label that read CORE1.DE-CIX.NET; a plaque that said MLX-32.

As airhead Jen on *The IT Crowd* asked: "This is the Internet? The *whole* Internet?" As machines go, I confess that it looked a lot like all the Internet's other machines. I had tried to prepare myself for this—for the possibility of banality, of an apparently unremarkable black box. This was like visiting Gettysburg: it's just a bunch of fields. The object in front of me was true and tangible, if unequivocally abstract; material, yet unknowable. I knew from Austin that this very machine was among the Internet's most important—the center of one of the biggest Internet exchanges—but it wore that significance discreetly. Its meaning had to come from inside *me*.

I was reminded of an extraordinary scene in Henry Adams's strange third-person autobiography, *The Education of Henry Adams,* when he describes his visit to the Great Exposition of 1900 in Paris. There he saw a miraculous new technology: a "dynamo," or electrical generator. It's a breathtaking confrontation with technology. Taken at face value "the dynamo itself was but an ingenious channel for conveying somewhere the heat latent in a few tons of poor coal hidden in a dirty engine-house carefully kept out of sight," he writes. Yet the dynamo meant everything to Adams: it became *"a symbol of infinity."* Standing beside it he felt its "moral force, much as the early Christians felt the cross." He continues: "The planet itself seemed less impressive, in its old-fashioned, deliberate, annual or daily revolution, than this huge wheel, revolving within arm's length at some vertiginous speed, and barely murmuring—scarcely humming an audible warning to stand a hair's-breadth further for respect of power." But it wasn't the machine's mystery or power that terrified Adams most. It was how clearly it signified a "break of continuity," as he puts it. The dynamo declared that his life had now been lived in two different ages, the ancient and the modern. It made the world new.

I felt the same way about this machine at the center of the Internet. I believe in the transformative power of the network. But I've always been at a loss for the physical symbols of that transformation. The Internet lacks monuments. The screen is an empty vessel, an absence not a presence. From the standpoint of a user, the object through which the Internet arrives is totally flexible—an iPhone, a BlackBerry, a laptop, or a television. But the DE-CIX was my dynamo—a symbol of infinity, pulsing with eight hundred billion bits of data per second. (*Eight hundred billion!*) It was louder, if smaller, than Adams's, and not on display in any grand hall but hidden behind half a dozen locked doors. But it was similarly a sign of the new millennium, something that made tangible the changes in society. I had come a long way from the squirrel in the backyard—and not only in miles traveled but in the movement from the edge of the network to the center, and in my understanding of the virtual world.

The guard tapped her foot impatiently. Nipper and I stepped around the back of the machine, where powerful fans blew away the heat generated by its efforts at directing all those bits, those fragments of each of us. The hot wind prickled my eyes, and I teared slightly. Then Nipper locked the cage, and we hustled out the doors.

In the car back to the DE-CIX office, Nipper asked if I was satisfied. Was it a good tour? I was looking for the *real* amid the merely virtual—something realer than pixels and bits—and I found that. Yet I was nagged by the thought that this machine in particular wasn't so different from so many other machines, which only seemed to reiterate the inconsequentialness of its unique reality. I believed in the importance of that particular box among others, but I felt far out on a limb in doing so. There were other boxes in the world, certainly. But equally, there was a world inside this box. I was far from done with the Internet. The essence of what I was after was the unique intersection of a place and a technology: a singular box in a singular city that stood as a physical crossroads of our virtual world. A bald fact of geography. Nipper and I drove along beneath the towering cranes of the port lands.

Back at the office, Orlowski fished around in a supply closet before emerging with my prize: a black T-shirt with large yellow writing on the front that said I ♥ PEERING. Then, out of another cardboard box, he produced a black Windbreaker, with a small DE-CIX logo on the breast. "Wear this in Amsterdam," Orlowski said with a wink. "And send my best to Job." Their competition extended to swag.

That evening in my hotel room I quickly typed up my notes and copied the audio files from my digital recorder onto my laptop. The hotel room desk faced the window. It had begun to rain and the rush-hour streets were noisy with traffic. A tram clattered by. Then, for added protection, and to soothe a lingering bit of paranoia, I copied everything to an online backup service, which I had learned was physically located in a data warehouse in Virginia, deliberately close to Ashburn. On my screen, I watched the status bar grow as the big audio files made their way toward it. I had a good idea which way the bits went. I was shading in the map.

After the stern grayness of Frankfurt, Amsterdam was a relief. I stepped off the tram into the middle of a busy spring evening on the Rembrandtplein, one of the living-room-like plazas that dot the center of the city. Pretty couples clattered by on heavy black bikes, coats floating open like wings. Forget the clichés of hash and whores; Amsterdam's liberalness seemed far deeper, more considered. Walking the quiet side streets lining the canals I looked through uncovered windows into living rooms lit by stylish modern chandeliers and filled with books. It reminded me of home in Brooklyn—or Breuckelen, as the Dutch called it, with our similarly tightly packed, *stoep*-fronted town houses.

In his book *The Island at the Center of the World,* Russell Shorto argues that this Dutch spirit is deep in the DNA of New York, and of America. There is, he writes, a shared "cultural sensibility that included a frank acceptance of differences and a belief that individual achievement matters more than birthright." It felt strange to apply these terms to the Internet—to think of it as something other than stateless, fluid, even postnational. The glass rectangles of our screens and the browser windows within them have had a flattening effect on the world greater than the presence of any McDonald's. Online, political borders are mostly invisible, smoothed over by the corporate triumvirate of Google, Apple, and Microsoft. But Amsterdam would begin to make the case otherwise. As it turned out, the Dutch Internet was very Dutch.

From early on, the Amsterdam Internet Exchange was heralded by the government as a "third harbor" for the Netherlands—a place for the bits, in the way that Rotterdam is a place for ships and Schiphol for planes. The Dutch saw the Internet as only the latest in a five-hundred-year lineage of technologies that could be exploited for national gain. "In the Netherlands, forts, canals, bridges, roads, and ports have always been first of military importance, and later very useful for trade," a 1997 op-ed argued. "The logistics of bits in the Netherlands will need a place of their own, in anticipation of catching a substantial proportion of the hundreds of billions of dollars at stake in the future world of Internet commerce." History had already proved the model: "Access to the open sea was in the time of the East India Company a decisive factor for success. . . . Providing access to the digital arteries of the global network will be decisive today." If Frankfurt had the luck of being in the geographic center of Europe, Amsterdam would have the pluck to make itself one of the logical centers of the Internet. If there was a broader lesson in this, it was the need for governments to invest in infrastructure—and then get out of the way. Throughout its history, the Internet has needed help getting going, only to benefit enormously when left alone.

The first ingredient in Amsterdam and everywhere was (and remains) fiber. In 1998, the Netherlands passed a law requiring all landowners to provide a right-of-way for private networks to lay fiber-optic cables—a right that was previously reserved for KPN, the national telephone company. Going one step further, the law stipulated that any company wishing to dig had to announce its intention and allow others to lay their own fiber and share the cost of construction. The intent was partly to keep the streets from being dug up repeatedly. But more important, the policy would eviscerate the old monopolies.

The results were successful, nearly comically so. I visited Kees Neggers, the director of SURFnet, the Dutch academic computer network, and a key player in the Internet's development in the Netherlands, at his office in a tower above the Utrecht railway station. From his bookshelf he pulled a bound report and opened it to a page of photographs taken at the moment of all that digging. In one, dozens of multicolored conduits bulged out of the soft ground of a polder, spreading out like a split whale on a beach. Another showed dozens of conduits pouring out of the doorway of an Amsterdam town house. Lain out on the cobblestone street, waiting to be buried, they filled a full lane of traffic. The colors—orange, red, green, blue, gray—indicated different owners; each contained hundreds of strands of fiber. It was an absurd abundance, and still is. "They shared the cost of digging along the dykes, toward the Amsterdam Internet Exchange, and that instantly provided an opportunity for all to connect—and to connect to each other very cheaply," Neggers said. "And then it grew like mushrooms."

I hadn't realized the extent of that growth until I stumbled across an online map, a homemade Google mash-up made by someone named only as "Jan," which indicated with colored pushpins, as if they were coffee shops, the location of nearly a hundred data centers in the Netherlands. A green pin meant an AMS-IX location, red showed private carrier-owned buildings, and blue indicated a data center that had fallen out of use. If I zoomed out to the scale of the country as a whole, the pins blanketed the screen, all leaning in the same direction, like windmills. It struck me as a startling example of the Netherlands's transparency: here, pulled together in one place on the open web, was the same information that WikiLeaks had deemed sensitive enough to bother leaking. And yet nobody seemed to care. The map had been there for two years already, apparently unmolested.

It also clarified a broader point that I'd been circling for months. It showed the Internet's small-scale geography, with the data centers clustered into defined Internet neighborhoods, like the industrial parks surrounding Schiphol Airport, the "Zuidoost" just southeast of the city center, and the academic area known as Science Park Amsterdam. A similar map for the area around Ashburn or of Silicon Valley would certainly show as many individual locations. But compared with those American-scaled suburban expanses the Netherlands was remarkably compact. It raised the possibility of a novel way of seeing the Internet, of gleaning its sense of place: a data center walking tour.

I was beginning to realize that I'd spent cumulative weeks behind electronically locked doors, engaged in long conversations with the people who made the Internet work. But these interactions had all been planned, considered, and tape-recorded. Permission had been granted by corporate higher-ups. Badges had been issued, visitor logs signed. But it often felt like I was wearing blinders. I had been so focused on the trees I had almost forgotten about the forest. I was always rushing through parking lots, diving headlong toward the "center." In nearly every case, the format of the day limited lingering. I had little time for any idle contemplation.

In Amsterdam, I was due to meet Witteman and visit the core switch at the heart of the Amsterdam Internet Exchange— the Dutch version of what I'd seen in Frankfurt. But the data center map seemed the perfect excuse for a more open-ended way of seeing the Internet, something more like a sojourn than another guided tour. The challenge was that the Internet was a difficult place to just show up. Data centers and exchange points don't have visitor centers, in the way of a famous dam or the Eiffel Tower. But in Amsterdam the Internet was so thick on the ground—there was so much of it—that even if I couldn't knock on doors and expect to be invited in, I could at least stroll by a couple dozen Internet buildings in a single afternoon, a piling up that would answer the question of what the Internet *looked* like in a new way. Architecture expresses ideas, even when architects aren't involved. What was the physical infrastructure of Amsterdam's Internet saying?

There's a wonderful essay by the artist Robert Smithson called "A Tour of the Monuments of Passaic." In his warped mind, the industrial wastelands of Passaic, New Jersey, become as evocative as Rome, with every inch worthy of aesthetic attention. But Smithson's insistence at playing the straight man ends up making the whole account wildly surreal. The New Jersey swamps become a place of wonder. "Outside the bus window a Howard Johnson's Motor Lodge flew by—a symphony in orange and blue," Smithson writes. The big industrial machines, quiet on that Saturday, are "prehistoric creatures trapped in the mud, or, better, extinct machines—mechanical dinosaurs stripped of their skin." His point is that there is value in noticing what we normally ignore, that there can be a kind of artistry in the found landscape, and its unconventional beauty can tell us something important about ourselves. I had the hunch this approach would work with the Internet, and with the help of that Dutch data center map I could set out to confirm it.

I enlisted a fellow wanderer—a professional Internet watcher, albeit of a more conventional stripe. Martin Brown worked for Renesys, the Internet routing table analysts, and he had recently moved to 's-Hertogenbosch, the small Dutch city known by most as Den Bosch, or "The Forest" (indeed!), where his wife had a new job with a pharmaceutical company. Brown, a former programmer and now a full-time watcher of the Internet's routing table, is an expert in the Internet's inner workings. In particular, I'd admired his study of Cogent and Sprint's "de-peering" event. Yet Brown said that while he'd been inside a handful of data centers and exchange points, he'd never stopped to really look at them—certainly not in this way at least. We made plans to meet for an urban hike, an eight-mile or so journey beginning at a subway stop a few minutes' ride from the city center and ending substantially farther out in the near suburbs.

Our first data center was visible from the elevated train platform: a menacing concrete bunker the size of a small office building, with worn-out blue window trim, spreading out along a canal connecting to the Amstel River. The late-winter day was gray and damp, and there were houseboats tied up at the edge of the still water. My map indicated that the building belonged to Verizon, but a sign on the door said MFS—the vestigial initials of Metropolitan Fiber Systems, the company that Steve Feldman worked for and that ran MAE-East, and which Verizon had acquired years before. There was clearly no rush to keep up appearances; it seemed, rather, that its new owners preferred the building to disappear. When Brown strolled up to the front door and peered into the darkened windows of the lobby, I nearly shouted at him, like a kid about to enter a haunted house. I admit I was skittish. The security barriers, frosted glass, and surveillance cameras made the point that this was not a place that welcomed outside interest, much less any visitors, and I was eager to avoid having to explain to anyone what it was exactly we were doing (much less what I had been doing all along), regardless of whatever map this place was on. The building said "back off."

We doubled back across the canal toward the data center that contained one of the AMS-IX cores. It was owned by euNetworks, which, like Equinix in Ashburn, is fundamentally in the business of renting space, and the building had a sign by the door and the friendly receptionist was visible through the lobby's glass walls. But when Brown and I walked around the back, its true starkness became clear: a block-long blank wall of gray brick at the base and corrugated steel above. Adding to the feeling of mystery was the rusting hulk of an old Citroën truck parked on the otherwise empty street, looking like a prop from *Mad Max*. We circled it, greedily taking pictures. (Smithson: "I took a few listless, entropic snapshots of that lustrous monument.") It had become an evocative scene: the lonely block, the too-blank brick and steel of the data center, the line of surveillance cameras, and—above all—the knowledge of what was happening inside those walls. This wasn't any old blank building but among the most important Internet buildings in the world. And this was getting fun.

We marched onward along a wide bike path, dodging middle schoolers pedaling home from soccer practice, and scampered across a couple wide intersections. After one false ID of an office building (too much glass to be a data center), we looped around an extra block to see an unmarked steel shed that seemed surprisingly sturdy. There were again no signs, but the map told me it belonged to Global Crossing, the big international backbone owner that has since been purchased by Level 3. Earlier in the week I'd been inside Global's facility in Frankfurt, and I could see the family resemblance. They were built at the same time, at the direction of a single engineer barnstorming across the continent. This was a piece of the Internet of the highest order, a key node of what Global liked to call its "WHIP," for "world's heartiest IP network." The corrugated steel and security cameras were the first clue, but more than that was the quality of the building's construction. A plumbing supply warehouse might be the same scale and the same materials, and might even have a camera or two. But this was definitely trying hard not to look like anything—the architectural equivalent of a police detective's generic sedan. Internet buildings are conspicuous in their quiet; but when you learn how to recognize them, they seem quietly conspicuous.

As we notched the miles, we learned their tells: the steel generator enclosures, the manholes out front forged with network names, the high-grade surveillance cameras. Trudging along was physically satisfying; we weren't sniffing the air with our smartphones looking for wireless signals, but divining more tangible clues.

We crossed beneath an elevated highway into a neighborhood of narrow lanes set alongside another canal, where there was a group of a half-dozen data centers interspersed among car dealerships. The data centers were bigger. A corrugated steel building with a narrow band of windows running the length of its second floor belonged (according to the map) to Equinix. Compared with the tilt-up concrete shells in Ashburn, it may as well have been designed by Le Corbusier, with its ribbon windows and paneled façade. But by any reasonable standard it was utterly unremarkable, a place we would have walked right by if we hadn't been walking straight toward it. We paused to admire it, and Brown sipped water from a canteen, as if we were on a mountain trail. A duck—green head, yellow beak, orange feet—waddled up beside us. We were chilled and wary. (Smithson: "I began to run out of film, and I was getting hungry.") I was tired enough at this point—not merely from our walk, but from a whole jet-lagged week in the stale air of the Internet—that the reality hit me hard: I was traveling the world looking at corrugated steel buildings. I had learned what the Internet looked like, generally speaking: a self-storage warehouse. An unusually pretty one, though.

The next day I visited Witteman at the AMS-IX offices. On the wall behind his desk was a homemade mash-up of the movie poster from *300,* based on the bloody comic book epic about the battle of Thermopylae. The original poster had read "Tonight

we dine in hell," and showed an enraged, bare-chested Spartan
baring his teeth. Witteman's version kept the soldier but Photoshopped the blood-dripping text to read, "We are the biggest!" I
had a hunch who in this fantasy represented the Persians. While
Frankfurt's exchange projected a polished character, AMS-IX
seemed to strive for a thoughtful informality, a philosophy that
extended to its offices in a matched pair of historic town houses
near the center of the old city. The young, international staff ate
lunch together every day, cooked by a housekeeper and served
family style at a table overlooking the back garden. There was a
homeyness to AMS-IX that I hadn't yet encountered in the Internet. Rather than the network being the realm of conspiracy theo-
ries and hidden infrastructure, the exchange embodied a spirit
of transparency and individual responsibility. And as it turned
out, that feeling extended to its physical infrastructure.

Before lunch, Witteman and I collected Hank Steenman,
AMS-IX's technology guru, from his office across the hall. The
three of us climbed into the AMS-IX jalopy, a beat-up little
minivan filled with old coffee cups, and headed toward the core
switch, located in one of the data centers Brown and I had walked
by. There was a bike rack outside and a welcoming, light-filled
lobby, with framed network maps on the walls. We walked down
a wide hallway lined with doorways painted bright yellow and
past a room used by KPN, filled with racks painted their patented
green color. AMS-IX had its own large cage in the back. The
yellow fiber-optic cables were perfectly coiled and bound. The
machine they plugged into looked familiar. Very familiar. It was
a Brocade MLX-32—the same as used in Frankfurt. Alas, the Internet's sense of place did not extend to the machinery. "So here's

the Internet!" Witteman teased. "Boxes like this. Yellow cable. Lots of blinking lights."

That evening, when I got back to the Rembrandtplein, a busker was singing like Bob Dylan, and tourists and revelers gathered around. Couples sat smoking on benches. A stag party stormed by, kicking up a commotion. Amsterdam was so many things. But all I could think about was what would happen if you sliced a section through the streets and buildings: the broken walls would glow with the prickly sparkle of all those serrated fiber-optic cables, another kind of red light: the rawest material of the Internet—and, even more than that, of the information age.

5

Cities of Light

Back in Austin, at NANOG, I'd met a guy named Greg Hankins with the unfortunate job title of "Solutioneer." He mixed with the peering crowd, was quick to pay for drinks, and seemed to be a member in good standing of the traveling circus of network engineers, peering coordinators, and Internet exchange operators. In particular, he was especially close with Witteman and Orlowski. But he didn't run a network or work for an Internet exchange. Hankins was employed by Brocade, a company that made—among just a few other things—the MLX series of routers. These were machines the size of refrigerators, the cost of trucks, and utterly essential for the inner workings of the Internet. In Frankfurt and Amsterdam I'd seen Brocade's most powerful model—the MLX-32—running full tilt. But I'd also seen it, or its less powerful versions, in nearly every other Internet building I'd been in, made by Brocade or one of its competitors, like Cisco or Force10. When I hadn't actually spotted the big router inside a locked cage, I saw its shipping box, littering the darkened corridors of the data center like the scat of a native bear. These were the cardboard shibboleths of the physical Internet, the clearest sign that a building was on the network in a serious way. But as much as that, I liked the way that routers were the basic building blocks of the Internet. They *scaled:* the twenty-dollar box I bought at Radio Shack was a kind of router, and so was Leonard Kleinrock's original IMP. They were and are the Internet's first physical pieces.

But what did I really know about what went on inside? I'd learned about the geography of the Internet, about where it was. But I didn't know much about *what* it was. At home, everything was copper: the wire coming from the backyard, the cables on my desk, the last vestigial telephone cords on the landline. But in the heart of the Internet, it was all fiber—thin glass strands filled with pulses of light. So far I'd been reassured that on the Internet there's always a distinct physical path, whether a single yellow fiber patch cord, an ocean-spanning undersea cable, or a bundle of fibers several-hundred thick. But whatever went on inside the router was invisible to the naked eye. What was the physical path in there? And what might that tell me about how everything else connected? What was the reductio ad absurdum of the tubes?

The Internet was a human construction, its tendrils spreading around the world. How was all that stuff shoehorned into what was out there already? Did it seep under buildings or along "telephone" poles? Did it take over old abandoned warehouses or form new urban neighborhoods? I didn't want a PhD in electrical engineering, but I hoped what was going on inside the black box and along the yellow wires could be ever so slightly, well, illuminated. Hankins was perpetually on the road and couldn't stop. But he had a guy in San Jose who could tell me something about the power of light.

Brocade's headquarters was in a mirror-windowed building in the shadow of the San Jose airport, in Silicon Valley. I was met in the lobby by Par Westesson, whose job is to string together Brocade's most powerful machines to simulate the largest Internet exchange points—and then to break them and figure out a way to make them better. "We'll pull out a fiber or power down one of the routers while traffic is flowing through," Westesson said. "That's a typical day for me." I got the impression he didn't like when things didn't work. Born in Sweden, he wore a neatly pressed checked tan shirt and brown chinos, and his blue eyes were dulled from time spent under fluorescent lights and amid the dry air of the laboratory upstairs. It was a room the size of a convenience store, busy with technicians standing in twos and threes in front of double-screened displays, or rustling through bins of fiber-optic cables and spare parts. The blinds were drawn against the sun. Westesson invited me to treat the place like a petting zoo.

I could take apart one of these machines—with no risk of harming the live Internet. The biggest and dumbest of a router's four basic parts is the "chassis," the file-cabinet-like enclosure that gives the machine its grossest physical structure, like the chassis of a car. Slightly smaller and smarter is the "backplane," which in an MLX-32 is a steel plate bigger than a pizza, etched with copper traces like a garden labyrinth. Fundamentally a router's job is to give directions, like a security guard in an office building lobby. A bit of data comes in, shows its destination to the guard, and says, "Where do I go?" The guard then points in the direction of the correct elevator or stairway, which is the backplane: the fixed paths between the router's entrances and exits. The third key element is the "line cards," which make the logical decision about which way a bit should go; they're like the security guard. Finally there are the "optical modules," which send and receive optical signals and translate them to and from electrical ones. A line card is really just a multiposition switch— hardly different from the input selector on a stereo. An optical module is a light—a bare bulb switching on and off. What makes it miraculous is its speed.

"So a gig is a billion," Westesson said, nonchalantly. He held in his palm an optical module of a type known as an SFP+, for "small form-factor pluggable." It looked like a pack of Wrigley's gum made of steel, felt as dense as lead, and cost as much as a laptop. Inside was a laser capable of blinking on and off ten billion times per second, sending light through an optical fiber. A "bit" is the basic unit of computing, a zero or a one, a yes or no. That pack of gum could process ten billion of them per second—ten gigabits of data. It inserted into a line card like a spark plug. Then the line card slid into the chassis like a tray of cookies into the oven. When "fully populated," an MLX-32 could hold well over a hundred optical modules. That means it could handle one hundred times ten billion, or a thousand billion bits per second—which adds up to the unit known as a "terabit," about what was flowing through that MLX-32 in Frankfurt on the Monday afternoon I visited.

"That's only where we are today," Westesson said. "The next step will be to start using one-hundred-gigabit-per-second links, where an individual fiber is transmitting at a rate of one hundred billion bits per second." They'd been testing them already. *Gig* is a word I encountered nearly every day, but it meant something different to try to count each bit separately.

Everything Westesson had been saying was about the "thickness" of the tubes: how much data is moving through the machine each second. But I was also interested in the corollary: What was the velocity of a single bit? This turned out to be something of a sore point. "Some of our customers are looking at how long it takes for a packet to get switched through a router. It tends to be in the microsecond range, which is one-millionth of a second," Westesson said. But compared with the amount of time it takes a bit to cross the continental United States, for example, that time spent crossing the router was an eternity. It was like walking ten minutes to the post office only to wait in line for seven days, around the clock. Brocade's machines, powerful though they may be, were the traffic-clogged cities on a journey across the open net. A millionth of a second was painfully slow, if that's possible to conceive.

According to the laws of physics, an unimpeded bit should be able to cross the three-foot cube router in five-billionths of a second, or five nanoseconds. Westesson showed me the math, jotting the numbers down on paper with a mechanical pencil: the speed of light through fiber divided by the size of the router. Then he checked his math using a calculator program on a nearby computer—which was kind of funny in itself, because of all the things we think about our computers being able to do, this kind of math was one of the last. He counted off the zeros on the screen. "This point is the millisecond . . . this point is the microsecond . . . and this one is usually expressed as nanoseconds, or billionths of a second." I mulled all the zeros on the screen for a moment. And when I looked up, everything was different. The cars rushing by outside on Highway 87 seemed filled with millions of computational processes per second—their radios, cell phones, watches, and GPSs buzzing inside of them. Everything around me looked alive in a new way: the desktop PCs, the LCD projector, the door locks, the fire alarms, and the desk lamps. The room had a watercooler with a green LED—and a circuit board inside! The air itself seemed electric, charged with billions of logical decisions per second. Everything in contemporary life is based on these processes, on this math. Only deep in the woods can we manage to turn them off, and then not entirely. Otherwise, in the city—forget it. The networked systems are everywhere: cell phones, streetlights, parking meters, ovens, hearing aids, light switches. But all invisible. To see it you had to imagine it, and in that moment I could.

But at this point Westesson was late for his next meeting, and becoming a little restless. I had the sense he wasn't late often. He walked me to the elevator. "Well, we've only just scratched the barest surface," he said. But it seemed we had actually gone quite far. In his essay "Nature," Ralph Waldo Emerson crosses "a bare common, in snow puddles, at twilight, under a clouded sky." And yet even that ho-hum journey brings him "a perfect exhilaration. I am glad to the brink of fear. . . . I become a transparent eye-ball; I am nothing; I see all." On a journey to the center of the Internet my bare common turned out to be the router lab. And what I saw was not the essence of the Internet but its quintessence—not the tubes, but the light.

And yet, where did that get me? Sitting on a bench outside taking notes, I wondered if this revelation demeaned my pilgrimage. After all, it came not at the building scale or even at the city scale, but at the nano scale. What if the Internet couldn't properly be understood as *places,* but was really better thought of as math made manifest; not hard, physical tubes, but ineffable, ethereal numbers? But by then it was time to get to the airport and I remembered that for all the constantly advancing miracles of silicon, the planet itself remains unassailable, along with the speed of light and the human desire to be connected. The bandwidth might expand, but California and New York and London do not get any closer together—all of which was painfully obvious on the long flight home to New York. The world is still large. In the taxi from JFK, inching through the familiar city streets, I was struck by how much of it there was. If the Internet was made of light, then what was all this other *stuff*—filling buildings, even whole neighborhoods, the whole glittering expanse of the skyline at night?

———

In December 2010, Google announced the purchase of 111 Eighth Avenue in Manhattan for $1.9 billion, making it the largest US real estate transaction of the year. It's a massive building, with nearly three million square feet of space spread across an entire city block, and it had been Google's New York City head-quarters since 2006. Google executives said the company needed the building to accommodate the expanding number of employees in the city. They already had two thousand people working there and were hiring like crazy. Owning the building outright would give them the flexibility they needed in the long term.

But Internet infrastructure people raised their eyebrows at that explanation. In addition to being prime office space in a popular neighborhood, 111 Eighth also happens to be among the most important network meeting points in the world, and certainly among the top three in New York. Google buying it was a little like American Airlines buying LaGuardia Airport—and claiming it only wanted it for the parking garage. Among many other companies, Equinix leased fifty-five thousand square feet of space in the building. But unlike in Ashburn or Palo Alto, so did many other data center companies and individual networks. At "one-eleven," as everyone calls it, the building itself was the exchange, with fiber running between individually leased spaces in complexly overlapping ways.

But what caught my eye about Google's purchase was a detail in a newspaper article, meant to explain Google's interest: 111 sat atop something called the "the Ninth Avenue fiber highway." Put that way, it sounded like the Hudson River, or maybe the Brooklyn-Queens Expressway. But when I started asking around, it turned out to not really be a thing, only a creative invention of real estate agents. Not that there wasn't a lot of fiber—there was tons. Just that it wasn't only under Ninth Avenue. New York City was filled with "fiber highways."

Walking around the city in my day-to-day life, I was captivated by the idea of the light pulsing beneath the streets. Climbing down the steps into the subway, I'd imagine the red lights sticking out of the concrete decking. This was the municipal corollary to what was going on inside the router. But it wasn't the realm of egghead engineers, their glasses reflecting with strings of numbers. It was about thick bundles of cable and dirty streets—an even heavier reality. I started to wonder just how that light got in the ground.

Hugh O'Kane Electric Company was founded in 1946 to maintain printing presses for publishers, but it had since evolved to become New York's dominant independent fiber-optic contractor. "We've got plenty of tubes around here," Victoria O'Kane, a granddaughter of the founders, said when I called. I wanted to actually see the fiber being put under the streets—the newest piece of Internet. Hugh O'Kane's crews did that practically every night. So one winter evening I rode the subway twenty minutes from home to a rendezvous on a downtown street corner with a white truck painted with lightning bolts.

On its bed was a spool of black cable the size of a Volkswagen. It was parked beside a manhole forged with the initials "ECS," short for Empire City Subway. But the "Subway" name wasn't what you might think. Empire City predated New York's "subway" transit system. Since 1891 ECS—now a wholly owned subsidiary of Verizon—had owned the franchise to build and maintain an underground system of conduits, which it offered for lease at published rates that haven't changed in a quarter century: a four-inch-diameter conduit will cost you $0.0924 per foot per month, while a two-inch one can be had for only $0.0578 a foot. Going the length of Manhattan will cost you about $4,000 a month—if there's space left in the conduits.

That evening, Brian Seales and Eddie Diaz, both members of Local Three, the International Brotherhood of Electrical Workers, were to install twelve hundred feet of new fiber under the streets, threading it through Empire City's existing tubes. The two worked for Hugh O'Kane, but the cable itself was owned by a company called Lightower and was extra thick: 288 individual fibers shoehorned into a package the diameter of a garden hose.

As they do most nights, Seales and Diaz left the garage in the Bronx at seven P.M. and "popped" the manhole at eight, heaving its 150-pound cover with steel hooks—together, as per union rules. The asphalt under my feet reverberated with the crash. The opened hole emitted a faint vapor that drifted across the shiny streets, glistening with the first light snowflakes of what would soon turn into a big storm. I was freezing. Seales was unbothered, his flannel shirt collar open. "I don't care how hard it snows, you can't get wet in a manhole," he said.

Toes to the curb, I leaned forward and looked inside the hole. There was no visible bottom, only an abyss of twisted cables. To give themselves more room to work, Diaz and Seales pulled out two big coil cases, rubber cans the size of Labradors marked "AT&T" and "Verizon," and laid them out on Broadway. They looked like giant squid under the streetlights, with their gray bodies dangling black cables. Some holes are so stuffed with cables that the cover pops right up, like snakes coming out of a can. The manhole was hard up against the security perimeter around the New York Stock Exchange. Bankers hustled by us, headed home from work. A cop inside a bulletproof hut cast a knowing eye in our direction. We were part of the nighttime rhythms of the city; after a day of moving money, the time had come to build and rebuild the city's more tangible pieces.

Seales had been working the streets of New York for Hugh O'Kane for sixteen years; for the eighteen years before that, he slung copper cables along the city's subway tracks. He looked like George Washington, with silver hair and a pointy nose. Diaz was younger and stockier, with dark hair and a twitchy face. On St. Patrick's Day, Seales calls him Eddie O'Diaz. Both wore their walkie-talkies clipped to the shoulder straps of their work bibs; to keep them from screeching when they stood too close to each other, each covered the speaker with his hand whenever his partner spoke, as if holding his hand to his heart.

The cable on the truck was a single continuous ribbon of fibers. An engineer sitting at a desk had drawn out the route on a big map of the neighborhood, indicating the path of the cable with a thick red line, and each manhole it traversed with a cowlick. In that form, there was nothing electronic about it. These were pure optical pathways, the least common denominator of the Internet. A fiber is a fiber—all they had to do was run it across the city.

The length being installed that night was what's known as a "lateral": a crosstown link connecting Lightower's two existing network spines, one heading up Broad Street and the other, Trinity Place. The immediate goal was to get 55 Broadway "on net" at the request of a single customer with (it seemed) heavy-duty data needs. Eventually this new stretch of fiber would also pick up additional customers along its path. It worked according to an incontrovertible physical truth: a pulse of light goes in one end and comes out the other. There is plenty of magic in the light itself—the rhythm and wavelength of its pulses determine the amount of data that can be transmitted at a time, which is in turn dependent on the machines installed on each end. But none of that changes the need for a continuous path. Individual strands of fiber can be spliced together end to end by melting the tips, like candles—but that process is delicate and time-consuming. The path of least resistance is unbroken. Hopefully.

The week before, Seales and Diaz had prepped the route. Using a fiberglass rod that folds into sections like a collapsible cane, they shoved a yellow nylon rope through the conduits and tied it off in each manhole along the way. Then they "dressed" the manholes, laying plastic tubing across each chasm to guide the cable. Tonight, they would pull the cable—twelve hundred feet total, just shy of a quarter mile—under the streets using the yellow rope. They'd start in the middle of the route, which also happened to be its highest point, the geologic spine of the island of Manhattan, Broadway.

Two other trucks would work with them, feeding the cable through the conduits, and pulling it out. When they were in position, Diaz hopped into the manhole. He was the "assist," the middle man in the bucket brigade. On the street, Seales wrapped the yellow leader rope around the truck's winch, and then fed the end down to Diaz. The cable would come out of the manhole, loop around the winch, and then go back in on its way to the next stop, where they would repeat the process. The truck idled heavily, its orange highway arrow illuminating the wet streets against the cycle of the traffic lights. When the call came over the radio—"Ready on the winch. GO GO GO"—Seales swung the broomstick-sized lever on the back of the truck and propped it in place with a plank. As the cable slid by, Seales greased it with a yellowish compound they call the "soap," which he sloshed out of a bucket with his hands. "Like K-Y, Astroglide, whatever," Seales said. "This stuff is dirty. It starts out white."

Diaz yelled up from the hole. "A couple Fridays ago, one of those nights it was in the teens, the shit was freezing on our gloves, on the wheel, it was cracking off the fiber as it was coming out. That was a night I wished I stayed in school. But I enjoy my job. I'm claustrophobic. I can't be in a building."

Up the block, the nose of the cable began to come out of the ground beside the other truck, half pulled and half pushed by Seales's winch. The guys there walked it into position with steady, rhythmic steps, crossing their legs and hinging their arms like doo-wop singers. As if doing a quarter-time square dance in the street, they laid the cable on top of itself in a figure-eight pattern. It looked like a woven basket the size of a hot tub.

"Some of these conduit runs are eighty or a hundred years old," Seales said. "They were put in when the city was built. Tonight we're in 2.5-inch iron ducts, which are very old, but down below there are square terra-cotta ducts that bricklayers put in in two-foot sections." The manholes are sometimes ornate, with arched spans. Seales can tell you a story about each one—like the "six-header" across from 32 Avenue of the Americas that's always filled with water. The morning of September 11, he was supposed to be pulling cable into the Twin Towers. Instead, he was in the basement of 75 Broad Street, pulling cable *from* them. It was his lucky choice. "The night before, I looked at the map, I looked at the route, and I said, 'If we get stuck late we're going to be out on the West Side Highway in the morning and the DOT is going to throw us off.'" So he turned the route around. When the Towers fell, "when all the shit was going on," he was on the other end of the cable, and his guys were safe nearby.

We rode in the truck to the next spot, two blocks away, bouncing leisurely up the middle of the empty street while the conduit underneath followed its own lopsided path. Diaz jumped out, and Seales positioned the truck so that its winch was directly over the manhole, ready to pull the fiber through. Its massive tire inched closer to the edge, and closer still, until I was sure he was going to fall in. "He's not going in there—that's a double wheel," Diaz assured. The union inspector was looking at his paperwork in the beam of the truck's headlights, and, as a gag, Seales tapped him gently with the bumper of his two-ton truck, like a priest giving communion. The inspector threw his papers into the air. Two women in high-heeled boots passed. "Whatta I got you for?" Diaz teased the inspector. "Not to watch out for me. That's two of them that went by."

When everyone was ready, the radio squawked: "GO GO GO."

Diaz yelled back: "GO GO GO." The winch ran smoothly for a few moments until the yellow rope snapped off the wheel. "Oh!" Diaz said. "That's not good." Work stopped as they searched for the problem. Somewhere under the street the cable had snagged.

"I prepped the route myself," Seales said in his defense. "I've worked these routes multiple times, with multiple customers." The problem turned out to be a "bareback" section—meaning the cable moved freely, rather than inside conduit or innerduct.

The joint between the fiber and yellow rope—known as the "nose"—had snagged. Diaz freed it up and called back over the radio, "GO GO GO." As the cable again slid by, Seales squinted through reading glasses to read the length, marked on the cable every two feet. "These fucking numbers are getting littler and littler," he said. The other truck, impatient to finish the night, sped up the winch, and Seales complained over the radio. "Slow it down slow it down." Getting no response, he yelled up the block. "Yo-HOOO! Nice and easy."

The cable pulled taut. Diaz coiled up an extra sixty feet, lassoed it with electrical tape, and banded it to the wall of the manhole—enough slack for the "splice truck" that would soon come to extract a couple fibers out of the cable and fuse them to another fiber coming out of the adjacent building. Seales stacked the orange cones, folded the steel safety gate that surrounded the manhole, and hooked the manhole cover back into place. It clanged, then thudded. "Another successful night," Seales said.

One evening a few weeks later, this new link would be "turned up." Its fibers would be spliced into their mates inside 55 Broadway's basement and hooked up to the proper light-emitting equipment—thereby increasing by the slightest increment the total accumulation of tiny illuminated tubes beneath lower Manhattan.

---

111 Eighth Avenue wasn't the only big Internet building in Manhattan, but it was the newest. The other two major ones—60 Hudson Street and 32 Avenue of the Americas—had a longer history as telecommunications hubs. But all three shared a defin-

ing characteristic: the fiber beneath the street was as important in their creation as the equipment in their towers. But the reason why had nothing to do with Google. It went back to a June night a hundred years ago.

"Without a single hitch the intricate task of transferring all the telegraph lines from the old building at 195 Broadway to its new quarters at 24 Walker Street was accomplished by the Western Union early yesterday morning," the *New York Times* reported on June 29, 1914. This brand-new operations building at the corner of Walker Street and Sixth Avenue—today known as 32 Avenue of the Americas—was to be shared between Western Union and AT&T, with AT&T occupying the first twelve floors and the top five belonging to Western Union. (It's worth pointing out here that the second "T" in "AT&T" stands for telegraph.) "When business is in full swing today 1,500 operators who have been working the sounding keys at 195 Broadway will be enjoying the conveniences of the most modern telegraph plant in the world," crowed the *Times*. By 1919, the building was among the nation's largest long-distance telephone central offices, with 1,470 switchboard test positions, 2,200 long-distance lines, and a transatlantic radio-telephone switchboard—all of which still wasn't sufficient to serve the country's telecommunications needs. Today, the building is one of the key pieces of New York's Internet—even if AT&T and Western Union's cohabitation didn't last.

In 1928, Western Union hired the architectural firm of Voorhees, Gmelin & Walker to design a brand-new twenty-four-story building of its own, three blocks south at 60 Hudson Street. Not to be shown up, AT&T hired the same architects to expand the old building to fill the entire block, as a new AT&T "long lines" headquarters. Undeterred by the stock market crash, the telecommunications rivals built matching art deco palaces, each with gymnasium, library, training school, even dormitories. The key to their separation lay beneath Church Street: an extensive run of clay conduits, filled with heavy-gauge copper wires that carried messages between the two systems—a sort of proto-Internet that would one day serve the real Internet. Both buildings were in full use until the 1960s, when the telegraph's decline eroded 60 Hudson's importance as a telecommunications hub.

But it wasn't the end for the building—and certainly not the end for the tubes under Church Street. Sixty Hudson's reinvention came with the deregulation of the phone industry, as AT&T's monopoly was chipped away by the federal courts. Western Union had vacated the building in 1973 but retained the rights to its "network"—most notably, the ducts leading back to AT&T. The former monopoly was being forced by the courts to allow competitors to connect to its system—but that didn't mean they had to provide the real estate to do it. It took William McGowan, founder and chairman of MCI—the fast-moving communications company that drove the fight for deregulation and would soon operate one of the first Internet backbones—to find a way. Discovering the unused conduits between the old buildings, he struck a deal for their use and established a beachhead inside 60 Hudson, with direct links to the basement of 32 Avenue of the Americas. The other competitive telephone carriers rushed into 60 Hudson after him, filling up the old telegraph floors one by one. Inevitably, those networks began to connect to one another inside the building, and 60 Hudson evolved into a hub. It's the paradox of the Internet again: the elimination of distance only happens if the networks are in the same place. "It's physical. It's proximity. It's the address," said Hunter Newby, an executive who'd helped make 60 Hudson a major Internet building.

Today, 60 Hudson is home to more than four hundred networks—that same number, and mostly the same networks, familiar from the other biggies. But a half dozen of these networks are of particular importance: the transatlantic undersea cables, which land at various points along the Long Island and New Jersey coasts, and then "home run" to 60 Hudson, where they connect with each other and everyone else. Astonishingly, the majority of them come from the exact same place: a building in London named Telehouse. Having so many of them in those two buildings wasn't planned and probably isn't prudent. But it does make sense, for the same reason international flights all land at JFK. "There's a recurring theme here: people go to where things are," Newby reminded me. Each network had its own equipment scattered around 60 Hudson in cages and suites of various sizes, but many of the ceiling-mounted fiber conduits come together at just a few places, known as meet-me rooms, operated by a company named Telx, a key competitor of Equinix. The largest meet-me room was on the ninth floor. It happened to have an excellent view, toward the AT&T building, four blocks uptown. Although the view wasn't the point. It was the path underground that mattered. These two buildings existed—they were the Internet—because of that link. I wanted to see it up close.

It was a hot day in the middle of summer when I met John Gilbert in the barrel-vaulted lobby of 32 Avenue of the Americas. Gilbert is chief operating officer of Rudin Management, the great family-owned New York City real estate company that in

1999 became only the second owner, after AT&T, of 32 Avenue of the Americas. He was an imposing figure in crisp white shirt-sleeves and silk power tie—a startling change from the network engineers in their hoodies. He stood beneath a lobby mosaic: an ocher-tinged Mercator projection beneath which is written the building's motto: "Telephone wires and radio unite to make neighbors of nations." "Why is radio in there?" Gilbert asked rhetorically, still gripping my hand. "When this building opened, there were no transatlantic telephone cables, only radios on buoys. Then, in 1955, *this* was built." He handed me a palm-sized copper cylinder, like a bloated penny, remarkably heavy and dense: a souvenir cut of the very first transatlantic telephone cable, called TAT-1, that connected the United States by wire to Europe for the first time. It ran from New York—from this building—to London, but the undersea portion itself stretched from Newfoundland to Oban, Scotland. Gilbert's grandfather designed it. As an engineer at Bell Labs, J. J. Gilbert wrote the specifications for "a submarine telephone cable with submerged repeaters." Gilbert kept the slice of cable on his desk, a totem to the physicality of telecommunications and his role as a keeper of it.

Since Rudin bought the place for $140 million Gilbert has been responsible for the building's continuing use as a communications hub, learning the peculiar needs of the business, and renovating the building to attract the new wave of Internet companies. In the beginning, his family connection to the building was a coincidence, but it soon fixed his role as the keeper of this place's history, from the telephone operators who first filled its floors to the huge fiber-optic distribution racks that do the same work today.

But in the decade after the Rudins bought it, 32 Avenue of the Americas evolved into a different animal from its sister building. At 60 Hudson, dozens of companies lease and sublease space for their equipment. But at 32 Avenue of the Americas, the Rudins both own the entire building and operate the telecommunications space, which they've named "The Hub." Elsewhere in the building are the offices of an architect, advertising agencies, and Cambridge University Press. But on the twenty-fourth floor is the Internet.

The "meet-me room" here was more of a "meet-me aisle," a single seventy-foot run of sixty-four racks, all filled floor to ceiling with looping yellow fibers, like a giant loom, accommodating tens of thousands of individual links. Gilbert led me gingerly around a maintenance guy high up on a ladder stringing new cables through the overhead trays—a more delicate process than I'd seen on the street outside. "This is your modern marketplace, where handoffs are made, fiber touches fiber, networks touch other networks," Gilbert said, as if showing off the marble bathroom of a Park Avenue co-op. In that way, the building isn't so different from Ashburn or Palo Alto—other than the fact that it is in this space that AT&T connected long-distance phone calls for half a century.

If Ashburn is a fluke of geography, this building is the opposite: a fact of geography. It was built upon hundred-year-old telephone infrastructure, nestled between stock exchanges and railroad tracks. It was wedged into the most natural piece of real estate at the elbow of the city's downtown and the first way out of town—the Holland Tunnel, to New Jersey and points west. And unlike the intentional sameness of the Equinixes of the digital world, its design is singular, its ways quirky and mysterious. It seemed to have evolved organically, as if compelled by its surroundings—feeding off its original root system of conduits and extending new ones over time.

Recently, a company named Azurro HD had moved in, taking advantage of the building's incredible abundance of bandwidth to help television broadcasters electronically transfer large quantities of video, rather than overnighting physical tapes. The company's small room was manned around the clock, and when we walked in to say hello, the technician on duty had a movie up on his enormous bank of mission-control-style screens: the 1975 espionage thriller *Three Days of the Condor*. Standing there inside one of the world's great "nexuses of information," we all watched for a long moment as the CIA agent played by Robert Redford tiptoed across the plaza of the World Trade Center towers.

Out in the elevator lobby, Gilbert opened a hinged steel door where the sliding elevator door should have been. Behind it was an open shaft crossed by a grated platform, with waist-high railings made of thin pipes. We stepped out onto it, so that below us were twenty-five stories of darkness—ignoring, of course, the hidden light inside the thousands of illuminated fibers. The wall of the shaft was lined with steel conduits and the plastic tubing known as innerduct—some orange, red, or dirty white, occasionally sliced open to reveal thick black cables tied neatly together in clumps. Networks that want to put in a new cable were required to protect it inside innerduct, but most opted for an extra layer of steel. The cables arched away from their vertical paths to join the ceiling fiber trays running above the data center space, as if a highway exit ramp were turned up into the air.

Gilbert and I then headed toward the basement—to the place where MCI breached AT&T's fortress. I lost track of how many doors we went through, but it was at least a half dozen by the time he moved an orange safety cone to the side, selected the right key from a massive chain, and opened an unmarked door. When the lights flickered on, I saw a grand room, like a walk-in closet for giants. The high ceiling ended in a ledge near the street wall, the kind of place a child might turn into a loft. The fiber-optic cables beneath the street traverse the building foundation through a special tube called the "point of entry." In the category of unique and expensive New York City real estate—$800 per month parking spaces, two-hundred-square-foot studios—these short pipes are high on the list of the strangest, and most expensive. In the early days of the big fiber-optic builds, meaning the mid-1990s, landlords hardly noticed them, approving requests to bring in new cables as needed. But as networks proliferated in buildings like this one, they increasingly clued into their value. Gilbert wouldn't specify, but I'd heard $100,000 a year wasn't unheard of—for a distance the span of your arms.

"When we first bought the building, the whole room was full of cables with these tags on them—*Des Moines*, *Chicago*—they went nonstop home run to those cities. We should have saved a couple of them," he said wistfully. Instead they employed three men to come in every day and cut away at the old cables, testing and checking each one to make sure it wasn't still filled with telephone calls before they removed it. It took the three of them two years to strip the room back to a state of emptiness, its cinder-block walls then repainted the same industrial gray paint as basements everywhere. Then the new fiber came.

I looked up at the spot where the ceiling met the wall. A massive twisted heap of black cables labeled with thick paper tags attached with twisted wires sprang down from above. There were steel cylinders and heavy-duty plastic fiber junction boxes, all tangled around one another like twine as they poured down from beneath the street. Some looped around each other, leaving bits of slack that could be sliced and spliced as needed. Other cables were thick and unbendable. Along one wall, vertical metal racks were installed to support additional cables that ran down the wall in neat rows, like garden hoses. If the equivalent room in Ashburn had all the intrinsic character of a mall bathroom, this room's strange shape revealed its long history, its building and rebuilding, the flitting ghosts of a century's phone calls, and the remnant of ten thousand nights of labor on the street above. It reminded me how much the Internet's physical presence was defined by the spaces between—whether inside the routers, or at the building point of entry.

I've been to a lot of secret places in New York, but few had that kind of presence. Partly it was the mysterious way we'd arrived there—past the subway entrance, up a few steps, down a few steps, through yet another door, one marked, another not, keys jangling, greeted by the rumble of a subway train and the flickering of the lights, then the slight tension in the air as Gilbert wondered what exactly it was I wanted to see here—and if it was really such a good idea showing it to me. But mostly what excited me was what I saw—or really imagined—in that huge cascade of thick cables: an incomprehensible lot of all the innumerable things we send over wires. Again, I realized that the words we use to describe "telecommunications" don't do justice to their current relevance to our lives, and certainly not to their corporeal presence. But it wasn't, again, a moment to linger. We had been in the room hardly forty-five seconds when Gilbert stepped back to the door and put his finger on the light switch. "So that's basically it," he said, and the light went out.

Considered at a certain scale, just beyond the foundation wall at 32 Avenue of the Americas were the conduits beneath the streets. It was one of those mysterious corners of Lower Manhattan that seem to peculiarly exist in the z-dimension—where crooked passageways led to barbershops, lost in both time and space. Following the tiled walls at odd angles, hearing a subway train just around the corner, you could never tell how deep you were in the bowels of the city, as if the world beneath the streets went on forever. But in the time you spent wandering, even if only for a minute, a much longer journey could have happened, many times over. Because looked at another way, just beyond the foundation wall at 32 Avenue of the Americas was London.

According to TeleGeography, the most heavily trafficked international Internet route is between New York and London, as if the cities were the two ends of the Internet's brightest tube of light. For the Internet, as for so much else, London is the hinge between east and west, the place where the networks reaching across the Atlantic link up with those extending from Europe, from Africa and India. A bit from Mumbai to Chicago will go through London and then New York, as will one from Madrid to São Paulo and Lagos to Dallas. The cities' enjoined gravity pulls in the light, as it pulls in so much else.

But despite that, the Internet's physical manifestation in the two cities is completely different. I had started out with the assumption that London is the old world and New York the new. But with the Internet, the opposite turned out to be true. If in Amsterdam the Internet was hidden away in low industrial buildings on the city's ragged edges, and in New York it colonized art deco palaces, in London it formed a single, concentrated, self-contained district—an office "estate," in the British term—just east of Canary Wharf and the City, known formally as East India Quay but by network engineers, and most everyone else, as just "Docklands." It was a massive agglomeration, an entire Internet neighborhood. I wondered what was at its heart. And how far into its center I could go.

Arriving in London a season later, I approached the district in suitably futuristic fashion, riding one of the driverless trains of the Docklands Light Railway system. It quickly left behind the elegant arched and tiled corridors of the old Underground and skimmed east behind the gleaming towers of Canary Wharf, with the names of the big international banks illuminated on their crowns. It was a kind of corporate utopia, an urban landscape lifted from the pages of a J. G. Ballard novel, "set in a mile-square area of abandoned dockland and warehousing along the north bank of the river"—that's from Ballard's *High Rise,* published in 1975. Its "high-rises stood on the eastern perimeter of the project, looking out across an ornamental lake—at present an empty concrete basin surrounded by parking-lots and construction equipment." And, "the massive scale of the glass and concrete architecture, and its striking situation on a bend of the river, sharply separated the development project from the run-down areas around it, decaying nineteenth-century terraced houses and empty factories already zoned for reclamation."

It wasn't a forgiving place. I repeatedly found myself standing on the wrong side of a high steel fence, staring up at an unseen guard through the blank glass eye of a surveillance camera, or resignedly boarding an empty London bus with a tap of my fare card—always pulled back into the System, always on the Grid. No doubt this is Ballard's world, albeit with a function beyond his imagining. East India Quay is iconically "super-modern," the double-edged term used to describe a landscape of sleek architecture and profound loneliness, ubiquitous surveillance cameras and lost souls. In *High Rise,* Ballard describes his protagonist's feeling of having "travelled forward fifty years in time, away from crowded streets, traffic hold-ups, and rush-hour journeys on the Underground," away from the dirty old metropolis to a tidier future. Ballard's description seemed so eerily prescient that it was hard to believe the area wasn't finished until nearly twenty years later. There's no mistaking that near futurism in the air of East India Quay—the distinct odorlessness of corporate control, the sense of a place defined by unseen forces. At every turn the place seemed eager to prove that reality is stranger than fiction. Or was it possible the reality was modeled on the fiction?

Across the Thames was the white moonscape roof of the Millennium Dome, sitting precisely on the prime meridian—like a cosmic affirmation of its importance. The East India station itself is a hundred yards short of the Eastern Hemisphere. The big Internet buildings line up along an empty plaza, looking like a showroom of giant chef's stoves, each steelier than the next. They have no signs, which is a shame given that their occupants' names are something Ballard might have invented: Global Crossing, Global Switch, Telehouse. I saw no pedestrians and little traffic, only the occasional white van marked with a telecom company logo, or a red double-decker bus idling at its terminus. The streets themselves take their names from the spices that filled the East India Company docks that once stood here: Nutmeg Lane, Rosemary Drive, Coriander Avenue. But the only tangible remnant of that past was a leftover piece of the brick wall that surrounded the docks. Beside a man-made pond overhung with weeping willows a bronze statue of two angelic figures struck a vacantly hopeful note—a *Winged Victory* of the data center.

The neighborhood was born as a network hub in 1990, when a consortium of Japanese banks opened Telehouse, a steely slab tower specially designed for their mainframe computers. An aerial photograph from the time shows it sprouting from a wasteland, with only the Financial Times Print Works next door (since turned into an Internet building). The banks came partly because the greater Docklands' status as an enterprise zone brought significant financial incentives, meant to spur its redevelopment following the departure of shipping to the deepwater ports down the Thames. But they really came for a more familiar reason: the site sat on top of London's main communications trunk routes, the fiber running like an underwater river beneath the A13 motorway. What was true in New York was true here as well: *people go to where things are.* And Telehouse was just getting going.

Almost as soon as the building was completed, the City was rocked by a series of terrorist bombings at the hands of the Irish Republican Army, leading the banks to scramble to install backup desks for "disaster recovery." Telehouse soon filled with empty trading halls, each desk a mirror to one in the City. Those first robust pieces of telecom infrastructure seeded the building for what came next: first, the deregulation of the British telecommunications system. And then the Internet. Being outside the influence of British Telecom made Telehouse the ideal place for the new competitive telephone companies to physically connect their networks. All those phone lines attracted one of the first British Internet providers, Pipex, which located its "modem pool" there: a few dozen book-sized boxes, bolted to a plywood frame and each linking a single telephone line into a shared data connection. Pipex was taking advantage of the building's telecommunications infrastructure coming and going, funneling the local phone lines into the international data links—which meant, at the time, a transatlantic circuit back to MAE-East. From that, the familiar recombinant growth of the physical Internet kicked in. The informal decisions of a handful of network engineers to build on its infrastructure had a profound impact on the future shape of the Internet.

Telehouse's position as the center got its semiofficial sanction in 1994, when the London Internet Exchange, or LINX, was established there, using a hub donated by Pipex, and installed next to its existing modem pool. At the time, a network could join the exchange only if it had "out-of-country" connectivity—which in practice meant its own link to the United States. The rule was famously snooty, enough to inspire the legendary quip that the LINX was run like a "gentleman's country club." But it did have an important, if unintended, consequence: the larger Internet providers began using Telehouse to resell their international connections to smaller providers. If you weren't a big enough network to cross the Atlantic yourself (and therefore be allowed to exchange traffic across the exchange), then you could at least connect to someone who did by renting a rack in Telehouse and installing your equipment. The last step was the most physical. "You could come into Telehouse and get the connection by dragging a piece of fiber across the floor," Nigel Titley, one of the LINX's founders, recalled to me. When before the end of 1994, BTNet—British Telecom's upstart Internet service—leased a two-megabit line across London and installed a router right next to Pipex's, it was clear that Telehouse had fully arrived. When new transatlantic fiber-optic cables began going in the water a few years later, it was obvious where they would terminate. The Eastern Hemisphere had a new center of the Internet. At Telehouse it all came together, an infinite mesh of small phone companies, commodities traders, pornographers, trading platforms, and website hosts, congealed into a global brain, with almost as many neurons.

Today everyone connected to the Internetworking community in London has a machine in Telehouse—and therefore a key. Almost every network engineer I contacted in London offered to show me inside. A gate slid open to allow in cars, but I entered through a full-body turnstile, unlocked by a guard watching closely from inside a booth. Telehouse had grown into a multi-building compound surrounded by a high steel fence. Security was intense. In 2007, Scotland Yard broke up an al-Qaeda plot to destroy the building, from the inside. Judging by the evidence gleaned from a series of hard drives seized in a raid on Islamic radicals, they were found to have conducted intense surveillance on Telehouse, as well as a complex of gas terminals on the North Sea. "Major colocation companies such as Telehouse are strategically important organizations at the heart of the internet," Telehouse's technical services director told the *Times* of London.

I crossed a narrow parking lot to a gleaming two-story reception hall, with glass walls on three sides and large ficus plants in the corners. I was met there by Colin Silcock, a young network engineer at the London Internet Exchange, who had offered to show me one of its cores—the descendant of Pipex's original box. We stepped into a pair of side-by-side glass tubes, each barely wide enough for a single person, with front and back doors that rotated open like a bank's security drum, and a wobbly rubber floor that floated free of the side walls—a scale that weighed you entering and exiting the building, to make sure you weren't leaving with any heavy (and expensive) equipment. As we stood there trapped for a long silent moment, waiting for the unseen computer to finish verifying our respective mass and identity, Silcock shot me a surprised look through the rounded glass. I had let out a burst of uncontrolled laughter, a loud snorting guffaw. I couldn't help it: I was inside a tube!

But as we went deeper into the building, Telehouse's high-tech bells and whistles dropped off, and a creakier reality came into focus. Where the neighborhood outside felt utterly controlled—scrubbed, anti-Dickensian—inside Telehouse the atmosphere tended to be more wayward. The original Telehouse building, now known as Telehouse North, had in the years since been joined by two new ones, each larger and more sophisticated than the first: Telehouse East, which opened in 1999, and Telehouse West, which opened in 2010. The three read like a short history of the Internet's architecture. The original owed a debt to the high-tech style made famous with the Pompidou Center. It had steel sun shades and a machinelike character. Inside, it was decidedly worn out, with shredded gray carpet, yellowing white walls, and huge bundles of unused copper cables flooding out of broken ceiling panels. The middle-aged building was neat and spare—inside, a study in linoleum. The youngest had a windowless façade patterned with steel panels, like pixels. It smelled of paint and spearmint, its rooms trafficked by technicians rolling dollies piled with brand-new equipment. A radiating system of catwalks and stairwells connected the buildings. It reminded me of a hospital, with heavy fire doors and archaeological layers of signage and building hardware scarring the walls. But instead of doctors and nurses, there were network technicians, nearly all men, studiously groomed with short cropped hair and goatees, looking like they might be about to leave for a nightclub, or perhaps had come straight from one. The parking lot exhibited their taste for tricked-out cars, and they carried bulky and unusual smartphones and large laptop backpacks. Nearly universally, they wore black T-shirts and zip-up hooded sweatshirts, handy for spending long hours on the hard floor of the server rooms, facing the dry exhaust blast of an enormous router.

As if going back in time, Silcock and I entered Telehouse North via a pedestrian bridge with an exposed sheet metal ceiling and scummy windows overlooking the car park. We followed the path of a ladder rack filled with purple cables—the only splash of color in the pallid environment. The hallway was littered with the cardboard box shibboleths and "caution" sawhorses laid across broken floor tiles. A guard sat in a straight-backed chair reading a paperback spy novel. Through a door window I saw empty desks, a remaining vestige of this building's role as a disaster recovery space. But most rooms were filled with aisles and aisles of high racks, stuffed with the same variety of equipment I saw in Palo Alto and Ashburn, Frankfurt and Amsterdam. In the corners were huge bundles of wires spilling from the ceiling, sinewy and broad like jungle tree trunks. Much of it was decommissioned; a popular joke has it that there's a fortune at Telehouse in copper mining. The streets outside were a rare moment when London felt empty, staid and binary; but this virtual world within seemed tossed and chaotic. It was a surprisingly shoddy piece of Internet. I understood what one engineer had meant when he described Telehouse North as "the Heathrow of Internet buildings." But a fantastically important one. The building's status as one of the most connected buildings on the planet forgave the broken pieces of flooring. At this point, it is what it is—and almost impossible to change. It would be like complaining that the streets of London were too narrow.

We finally arrived at the London Internet Exchange's hotel-room-sized space, cluttered and homey, crowded with the detritus of the long hours the engineers spend in there. Blue Ethernet cables hung like neckties on a series of hooks, alongside the coats. Silcock gave me a little tour, identifying the different pieces of equipment and filling in some of the history of the exchange. It was getting to be lunchtime, I was hungry, and I almost left without noticing it. But tucked away at the end of a narrow aisle of equipment, blinking innocently away, was another of those refrigerator-sized machines: a Brocade MLX-32, from a mirror-walled building in San Jose, California. Silcock propped his laptop on a chest of tools and looked up its live traffic numbers. Moving across the switch at that instant were three hundred gigabits of data per second, out of a total of eight hundred gigabits across the London Internet Exchange as a whole. Deep in the heart of Telehouse I could hear Par Westesson's voice in my ear, as clear as if he were on the telephone. *A gig is a billion,* he said. A billion bits made of light.

6

The Longest Tubes

The underwater telecommunications cable known as SAT-3 sweeps down the Atlantic coast of Africa from the western edge of Europe, linking Lisbon, Portugal, to Cape Town, South Africa, with stops along the way in Dakar, Accra, Lagos, and other West African cities. When it was completed in 2001, it became the most important link for South Africa's five million Internet users, but a horribly insufficient one. SAT-3 was a relatively low-capacity cable with only four strands of fiber, while the biggest long-distance undersea cables might have up to sixteen. Worse, its meager capabilities were further reduced by the needs of the eight countries SAT-3 connected before arriving in Cape Town. South Africa was the bandwidth equivalent of an attic shower. The country faced a widely discussed "bandwidth crisis," with low usage caps and exorbitantly high prices.

This troubled Andrew Alston more than anyone. As chief technology officer of TENET, South Africa's university research network, Alston had been a slave to SAT-3 since its completion, purchasing ever-larger amounts of bandwidth to serve the growing needs of the entire academic system. By 2009, Alston was paying nearly $6 million a year for a 250-megabit connection.

Then a new cable, SEACOM, arrived. It ran up the eastern coast of Africa, stopping in Kenya, Madagascar, Mozambique, and Tanzania, before branching to Mumbai and through the Suez toward Marseille. Alston signed on as a charter customer with a ten-gigabit connection—forty times the bandwidth he had on SAT-3, at the same price. But the "circuit" had very specific geographic terms: it linked the cable's landing point in the coastal village of Mtunzini, ninety miles from Durban, straight to Telehouse in London, where TENET had existing connections to more than a hundred other networks. That left Alston to complete the final link between Mtunzini and Durban, where his nearest router lived. Cabling everything up and configuring the fiber-optic equipment took forty straight hours. At the end of it, he was sitting cross-legged and a little delirious on the floor beside his equipment when a light indicated that the connection was active—the whole ten thousand miles to London. "This was probably 4:30-odd in the afternoon, and—pujah!—I could see both ends of the link," he recalled. He tried to run a few tests but quickly maxed out his equipment; the capacity far exceeded what his computer could artificially generate. He had his finger on a gusher.

He told me this story over the phone from his office in Durban, while I sat in my own office in Brooklyn. The phone line was crisp, the two hemispheres and fifteen thousand or so miles of cable between us amounting to only the slightest discernible delay. But I was aware enough of the distance to be even more shocked by the bald physicality of what he had described. We all deal constantly with the abstraction of an Internet connection that's "fast" or "slow." But for Alston the acceleration came with the arrival of an unfathomably long and skinny thing, a singular path across the bottom of the sea. Undersea cables are the ultimate totems of our physical connections. If the Internet is a global phenomenon, it's because there are tubes underneath the ocean. They are the fundamental medium of the global village.

The fiber-optic technology is fantastically complex and dependent on the latest materials and computing technology. Yet the basic principle of the cables is shockingly simple: light goes in on one shore of the ocean and comes out on the other. Undersea cables are straightforward containers for light, as a subway tunnel is for trains. At each end of the cable is a landing station, around the size of a large house, often tucked away inconspicuously in a quiet seaside neighborhood. It's a lighthouse; its fundamental purpose is to illuminate the fiber-optic strands. To make the light travel enormous distances, thousands of volts of electricity are sent through the cable's copper sleeve to power repeaters, each the size and roughly the shape of a bluefin tuna. One rests on the ocean floor every fifty or so miles. Inside its pressurized case is a miniature racetrack of the element erbium, which, when energized, gooses along the photons, like a waterwheel.

It all struck me as wonderfully poetic, an ultimate enjoining of the unfathomable mysteries of the digital world with the even more unfathomable mysteries of the oceans. But with a fun-house twist: for all the expanse these cables spanned, they were skinny little buggers. There wasn't all that much to them. The cables spanned oceans and then landed at incredibly specific points, tying up to a concrete foundation inside a manhole near the beach—a far more human-scaled construction. I imagined them like elevators to the moon, diaphanous threads disappearing to infinity. In their continental scale, they invoked *The Great Gatsby*'s image of an expanse "commensurate to [man's] capacity for wonder." Our encounters with this kind of geography typically come with more familiar images, like a ribbon of interstate, a length of train track, or a 747 parked expectantly at the airport gate. But undersea cables are invisible. They feel more like rivers than paths, containing a continuous flow of energy rather than the occasional passing conveyance. If the first step in visiting the Internet was to imagine it, then undersea cables always struck me as its most magical places. And only more so when I realized their paths were often ancient. With few exceptions, undersea cables land in or near classic port cities, places like Lisbon, Marseille, Hong Kong, Singapore, New York, Alexandria, Mumbai, Cyprus, or Mombasa. On a daily basis it may feel as if the Internet has changed our sense of the world; but undersea cables showed how that new geography was traced entirely upon the outlines of the old.

For all that magic, my journey to see the cables began in an office park in southern New Jersey. The building was true Internetland—unmarked, shiny, near the edge of the highway, with apparently no one around except the FedEx guy. It belonged to Tata Communications, the telecom wing of the Indian industrial conglomerate, which in recent years had made a strong push to be a major competitor among the Internet's global backbones. In 2004, Tata paid $130 million for the Tyco Global Network, which included almost forty thousand miles of fiber-optic cable spanning three continents, including major undersea links across both the Atlantic and Pacific Oceans. The system was a beast. Tyco was best known for manufacturing cables, not owning them, but as part of the corporate largesse-turned-malfeasance under CEO Dennis Kozlowski—who was convicted of grand larceny and securities fraud and sent to prison in 2005—Tyco spent more than $2 billion building a global network of its own, at an unprecedented scale. The piece of the network known as TGN-Pacific, for example, consisted of a fourteen-thousand-mile loop from Los Angeles to Japan and back to Oregon—two full crossings of the mighty Pacific. Finished in 2002, it had eight fiber pairs, double the number of its competitors. From an engineering standpoint, the Tyco Global Network—rechristened the Tata Global Network—was grand and beautiful. But financially the project was an unmitigated disaster, perfectly timed for the 2003 low point of the technology industry. As the Englishmen who dominate the undersea cable industry liked to say, the capacity they're selling is too often "cheap as chips."

Simon Cooper was Tata's Englishman, with the job of making the company's investment pay off. Internet traffic has grown continually in the last decade, but prices have fallen just as fast. Tata planned to buck the trend by finding the places in the world with latent potential. Its strategy was to be the telecommunications network that finally linked the "global south," the poorer—

and less connected—regions of the world, especially Africa and South Asia. Cooper spent his time trying to decide what countries to plug in next. Recently, he'd begun an ambitious building program to supplement the original Tyco network with even more cables—stringing them around the earth like lights around a Christmas tree.

In New Jersey, I waited a few minutes in the office kitchen, watching a group of Indian engineers make tea. Then, just before ten o'clock, I was invited into a conference room dominated by three giant flat-screen televisions, lined end to end facing a long table. Cooper was sitting inside the middle screen. He was in his early forties, with a shiny pate and a cheerful grin, looking a little worn out, alone in a room in Singapore late at night, coming to me via Tata's high-end video-conferencing link. We'd spoken once before. That time, Cooper was in an airport lounge in Dubai at midnight. He seemed to be constantly roaming, physically and mentally, as if he were the human incarnation of the network itself. I suppose the fact that I was talking to a TV had something to do with it, but I couldn't shake the notion of Cooper as a man inside the Internet. In a business filled with obfuscation, he was good humored and direct. I knew why: Tata was eager to compete with the AT&Ts and Verizons of the world, which meant improving their name recognition in the United States—and inviting over any journalists who asked.

"We've done the belt around the world and now we're reaching up and down a little bit," Cooper said nonchalantly, talking about the planet as if it were his lawn. Tata had extended its cable between the United States and Japan with a new link to Singapore and then onward to Chennai. Then from Mumbai another

Tata cable passed through the Suez to Marseille. From there, the routes went overland to London, and finally connected to the original transatlantic cable that connected Bristol, England, to New Jersey. Cooper made it sound like no big deal, but he'd built a beam of light around the world.

To go "up and down," Tata bought a stake in SEACOM, the new cable to South Africa, as well as another new cable down Africa's western coast, intended to break SAT-3's grip. And they were making a move into the Persian Gulf, planning a new cable that would connect Mumbai to Fujairah on the Emirates' eastern shore, and then around the Strait of Hormuz to Qatar, Bahrain, Oman, and Saudi Arabia. The cable would go from port to port around the gulf like a packet ship.

"You get a number of benefits from being global," Cooper said from inside the screen, chopping at the desk on the other side of the world. "We're connected to thirty-five of the biggest Internet exchanges around the world, so you can get to DE-CIX or AMS-IX or London, whether it's the last mile, or the last three thousand kilometers. And we get to talk about our global restoration, our round-the-world capability." In other words, Tata could promise that if its path from Tokyo to California were somehow obstructed—by an earthquake, say—they'd happily send your bits around the other way. It reminded me of Singapore Airlines' two daily flights from New York to Singapore: one goes east and one goes west. But only with the Internet do we treat the scale of the planet so casually—and only then because we have physical links like these.

For Tata, it was all an effort toward connecting the unconnected places—and thereby getting away from the falling prices on the generally overserved routes across the Atlantic and Pacific. "Look at Kenya," Cooper said. "Last August it had only satellite. Suddenly it's as well served as most other coastlines around the world, with the exception of hot spots like Hong Kong that have ten or twelve cables. But it's gone from zero to three cables in eighteen months. That makes it part of the global network. Not every customer wants a link from Kenya to London, but once you can do it, and do it consistently and do it well, people begin to think about things like call centers, which are constantly hunting for the place with the lowest cost services. The demand springs up."

Undersea cables link people—in rich nations, first—but the earth itself always stands in the way. To determine the route of an undersea cable requires navigating a maze of economics, geopolitics, and topography. For example, the curvature of the planet makes the shortest distance between Japan and the United States a northern arc paralleling the coast of Alaska and landing near Seattle. But Los Angeles has traditionally been the bigger producer and consumer of bandwidth, exerting a southward tug on earlier cables. With TGN-Pacific, Tyco solved the problem the expensive way, by building both.

Further complicating the geography is the demand for low "latency," the networking term for how long it takes information to travel across the cable. Latency used to be a concern only for the strictly telephone people, eager to avoid an unnatural delay in conversations. But more recently it's become an obsession of the financial industry, to serve the needs of high-speed automated trading, where computers arbitrage based on knowing the market news an extra millisecond in advance. Since the speed of light through a cable is consistent, the difference is entirely in the length of its path. Tata's route from Singapore to Japan is more direct than its competitors', which also gives it the fastest travel times all the way to India. But Tata's transatlantic cable is frustratingly slow. Tyco originally connected it to a landing station in New Jersey, close to its corporate headquarters. But compared with the transatlantic cables that landed on Long Island, by the time a bit went down the coast and back up to the city, the route effectively made London and New York two hundred more miles apart. At the time no one thought it mattered. "Now I get beaten up in meetings because there's one millisecond extra compared to our competitors," Cooper said, rubbing his brow. The first new transatlantic cable in a decade will be laid in 2012 by a small company called Hibernia-Atlantic. They designed it from scratch to be the fastest.

The micro geography matters just as much. Specialized ships conduct surveys of the ocean bottom, carefully plotting routes over and around underwater mountains—like grading a railroad, but without the option to dig any tunnels. The paths carefully avoid major shipping lanes, to limit the risk of damage from dragging anchors. Because if a cable does fail, a repair ship is dispatched to lift both ends to the surface using grappling hooks and fuse the ends back together—an expensive, slow process. Occasionally, the situation becomes more dramatic.

All but a few cables between Japan and the rest of Asia pass through the Luzon Strait, south of Taiwan. Looking at the map it's easy to see why: routing south around the Philippines would add too much mileage—and therefore cost and latency. But the Formosa Strait between Taiwan and mainland China is dangerously shallow, putting any cable at risk of being struck by fishermen. What's left is the Bashi Channel in the Luzon Strait, which with depths of up to four thousand meters seemed the perfect cable highway.

Perfect, that is, until Boxing Day 2006, when just after eight in the evening local time a 7.1 magnitude earthquake struck south of Taiwan, causing major underwater landslides that severed seven of the nine cables passing through the strait, some in multiple places. More than six hundred gigabits of capacity were knocked offline, and Taiwan, Hong Kong, China, and most of South Asia were temporarily disconnected from the global Internet. Taiwan's Chunghwa Telecom reported that 98 percent of its capacity with Malaysia, Singapore, Thailand, and Hong Kong was offline. The big networks scrambled to reroute their traffic on the working cables, or send it around the world the other way. But trading of the Korean won was temporarily halted, an Internet service provider in the United States noticed a sharp decrease in Asian-born spam, and a provider in Hong Kong publicly apologized for YouTube's slow speed—a full week later. It was two months before things were back to normal. And the name "Luzon" still brings shudders to network engineers.

At the logical level, the Internet is self-healing. Routers automatically seek out the best routes among themselves. But that works only if there are routes to be found. At the level of physical cables, rerouting traffic means creating a new physical path, stringing a fresh yellow patch cable from the cage of one network to the cage of another—maybe in Equinix's facility in Tokyo, or at the Palo Alto Internet Exchange, or inside One Wilshire in Los Angeles, all of which are places where the major transpacific networks have meeting points. But otherwise network owners are faced with the excruciatingly analog task of snatching up cables from the ocean floor with steel hooks. After Luzon, Tata had three ships in the area for nearly three months, picking up cables, splicing them together, lowering them back down, and then moving on to the next break. So when Tata planned a new cable in the region—the first postearthquake— Cooper thought twice about the route. "We went as far south as we could, which maybe isn't the optimal route from Singapore to Japan, but if there's an earthquake in the old place, we won't be affected, and if there's an earthquake near us, the other networks will generally stay up," he told me, sitting up in his chair in Singapore, which matched mine in New Jersey. "You make these tactical decisions." And then you turn the decisions back to the economics. Vietnam has eighty million people and poor connectivity. "Maybe they'd be interested in a new cable?" Cooper ventured. I tried to imagine what that would be like, a new cable hauled up on a white sand beach in Vietnam. Of all the moments of the Internet's construction, it struck me as the most dramatic—the literal plugging in of a continent. I asked Cooper if Tata might have a new cable landing anytime soon. With enough warning, and if they didn't mind, I'd try to be there to see it.

"Actually we do have a landing coming up," he said from inside the TV.

"Where!?" I shouted. Then I got worried. What if it was on the other side of the world, maybe in Guam (a big cable hub) or Vietnam? Or what if it was somewhere not entirely conducive to visiting journalists watching critical infrastructure, like Bahrain, or Somalia? This might not be so easy. But Cooper was cool. "It depends on the weather," he said. "We'll let you know."

In the meantime, I set out for the spiritual home of undersea cables. If the Internet's newest links tended to settle in the corners of the map, the old ones concentrated in more familiar places, and in one place above all others: a small cove called Porthcurno, in Cornwall, near the western tip of England, just a few miles from Land's End. Throughout the entire 150-year history of underwater communications cables, Porthcurno has been an important landing spot, but also a training ground—the cable world's Oxford and Cambridge. Looking at a map, I could easily see why. The geography hadn't changed. Land's End was still the westernmost point of England; England was still a hub for the world. According to TeleGeography, the busiest intercontinental route is between New York and London—primarily from 60 Hudson to Telehouse North. Several of the most important physical paths passed through Porthcurno.

But visiting a cable landing station wasn't as easy as getting inside the big urban hubs. The Docklands, Ashburn, and others had a constant stream of visitors. Security was tight, but there was a sense of them as inherently shared places, nearly public ones. But the cable landing stations were quietly hidden away, and they rarely received visitors. But Global Crossing, then the operator of a major transatlantic cable known as Atlantic Crossing-1, finally responded to my entreaties—perhaps officials were pleased I was paying attention to something other than the company's spectacular 2002 bankruptcy. My press contact only asked that I have a chat with her director of security, who would in turn "notify his government contacts" of my plans. Ah yes, *those* plans: to visit the Internet.

A little while later I was boarding a Penzance-bound train at London's Paddington Station. The iron arches of its rail shed were the perfect send-off. Paddington was designed by Isambard Kingdom Brunel, the greatest of Victorian engineers, who also laid the route and surveyed the tracks of the Great Western Railway, toward Bristol and beyond. But Kingdom Brunel also designed the SS *Great Eastern,* the largest ship in the world at the time of her completion in 1858, specifically designed to carry enough coal to steam to Trincomalee, Ceylon (today's Sri Lanka), and back, a distance of twenty-two thousand miles. He and Simon Cooper would have had something to talk about, even more so given the *Great Eastern*'s most famous use: the laying of the first transatlantic telegraph cable, whose 2,700-mile length could be coiled in the massive ship's hull. Cooper would have especially liked the early transmission rates: $10 per word, with a ten-word minimum. Practically speaking, I was on my way to Porthcurno. But I was aware that really I was on the trail of a broader idea about the triumph of technology over space—and for that there was no better patron saint than Kingdom Brunel.

Within a few hours the tracks looked out over the stormy seas where the English Channel met the Atlantic. Britain was beginning to feel like the island it is. With each mile the view out the window became more nautical. I was headed to the end of the gangplank, a spit of land known as the Penwith Peninsula—the westernmost of the pincers that look as if they're about to snip at the ships entering the Channel. To my American eyes the landscape was ancient, with cragged trees, roads that sat deep in the fields, and stone farm buildings that seemed to be sinking. Penzance was the end of the line. The beach concessions were all closed for the season but the "Prom"—the seaside promenade— was busy with walkers along the sweeping bay. I rented a car at the station, and since it was midafternoon in fall and I had no appointments for the afternoon, I decided not to bother with a map and aimed only for the sun, feeling my way toward Porthcurno. I figured it would be hard to get lost. There was only one way to go. I had reached the end of the land.

Porthcurno nestles at the base of a valley, a few dozen tidy houses huddled along a narrow lane that ended at a spectacular beach, a short crescent beneath high cliffs. The vegetation was nearly tropical, with scrub trees and flowers, and the water was turquoise. The Falmouth, Gibraltar, and Malta Telegraph Company landed its first cable here, to Malta, in 1870. The beach was chosen over Falmouth, forty miles east, after concerns that the cable there could be damaged by anchors in the busy port. (Cooper would have done the same thing.) Within a few years, two hundred thousand words were transmitting by telegram through Porthcurno annually, and new cables were planned. By 1900, Porthcurno was the hub of a global telegraph network that linked India, North and South America, South Africa, and Australia. By 1918, 180 million words were passing through the valley annually. At the start of World War II, Porthcurno—or "PK," in telegraph notation, a reference to its original name of Porth Kernow—was the largest cable station in the world. The company by then known as Cable & Wireless operated fourteen cables from the valley, totaling 150,000 miles in length. To protect them from Nazi sabotage, flamethrowers were installed on the beach, and miners were brought in to hollow out the granite hillside and move the station underground. After the war, Cable & Wireless took over the expanded facilities as its training college. Employees from all over the world converged in the valley for classes, to learn how to run the equipment and the business, before being posted to Cable & Wireless stations overseas. The school was active as late as 1993, seeding a close fraternity of men who still fondly recall their days in Cornwall. Porthcurno is their spiritual home, and today the bunker houses the Porthcurno Telegraph Museum, where much of the original equipment is on display and history videos play on a loop.

That evening I was one of two diners at the Cable Station Inn, the pub that occupies the training college's old recreation center, purchased by its proprietors directly from Cable & Wireless. My journey didn't seem strange to them. Their neighbor— and a good customer, it sounded—managed one of the landing stations and was a bit of a know-it-all. "He'll talk your ear off, explaining everything, but he knows more than anyone about this sort of thing," the publican told me. "Google googles him!"

"Maybe he can give you a tour?" his wife ventured.

"No, it's not such an easy thing," he corrected her.

I visited the archives at the Telegraph Museum the next morning. A pensioner working her way through a stack of Porthcurno's old school registries let out a shriek: her uncle was born before her grandparents married. I sat at a long wooden table in the old college building as Alan Renton, the archivist, pulled out boxes of documents from early cable landings on the beach and surveyors' maps of the bay. The engineer's report of the "Porthcurnow—Gibraltar No. 4 Cable," lain in 1919, was a testament to competence if ever there was one. The cable ship *Stephan* left Greenwich with 1416.064 "nauts"—nautical miles—of cable, made by Siemens Brothers, late in November. A few days later, in a gentle northeasterly breeze, she anchored in the cove at "PK" and sent the end of the cable ashore, supported in the water by ninety wooden casks. By 5:20 that evening, the anchor was hoved in, the cable was paying out over the stern, and the *Stephan* was steaming toward Gibraltar, "all proceeding satisfactorily." Within two weeks she was in Gibraltar Bay, preparing to land the other end of the cable on a "fine, bright and clear" day. "Completed final tests and advised the managing Director," the report concluded. Cable laying was already routine (notwithstanding the engineer's complaints about "the obvious risks of laying cables in deep water in Winter time on crowded seas and the fact that the *Stephan* is difficult to handle"). It was a reminder that Porthcurno had already been the communications center of a humming empire for two generations then—and would be for a long time to come, if more quietly so.

Late that afternoon I walked down to the beach where the old telegraph hut was maintained by the museum and opened up for visitors on good beach days. The sun was setting against the cliffs, and there were only a few couples staring out at the water. High up on the beach was a worn-out sign that said TELE-PHONE CABLE, as a warning for passing boats. I hiked up a steep staircase built into the rocks to a path along the cliffs. A fishing boat passed far below, a speck smaller than my fingernail.

Way out at sea, a big tanker steamed toward the Channel. The ocean was a flat steel-blue carpet stretching to the horizon, an image of infinity. I tried to picture the cables on the ocean floor, in their last few feet before landfall. In the museum's gift shop I'd bought a small sample of actual cable, mounted in a vitrine the size of my thumb. The cable's plastic casing was cut away to reveal the power-conducting copper tube and the fibers within. It was smaller in diameter than a quarter—but went on forever. The whole thing was simultaneously accessible and inaccessible, easy to grasp in one dimension but hardly imaginable in the other. It was like the ocean itself: the biggest thing on earth, yet traversable by plane in a day or electronically in an instant. How strange to be reminded while looking for the Internet, so often rhapsodized as making the world smaller, just how big the world is. The network hadn't erased distance, but left its streaks visible, as if on a just-cleaned blackboard.

Walking back toward the village, I saw a manhole with the word *ductile* forged into it. As I approached the beach parking lot, there were more manholes and then a little compound of equipment surrounded by a wood fence, nestled in the reeds. It hummed. Sprouting out of a drainage ditch were huge prehistoric stalks of *Gunnera,* or giant rhubarb, each one bigger than a man—as if their growth were nurtured from below by the light passing beneath them.

That night at the B&B, I Skyped with my wife in New York, about the drawings our daughter made at day care, the mess the dog made, the man who was coming to fix the leak. Unlike a phone call, our conversation went over the Internet; it was free and crystal clear, composed of something like 128,000 bits each second. Afterward, out of curiosity, I ran a traceroute to see if I could discern which way they had all gone. The path went back to London—before doubling back through here to New York. The B&B was perched nearly on top of the road, and beneath the road was an umbilical connecting the United States and Europe. But it flew right by without stopping, like the jetliners high above. When I turned out the light, the valley was so quiet that my ears rang.

The next morning, the Global Crossing station manager, Jol Paling, met me at the B&B and I followed him in my car toward the landing stations. Immediately out of the valley we came upon what amounted to a High Street of the undersea cable world—a half-dozen stations lined up along the road. The first was disguised as a stone house and would have been unrecognizable as a landing station at all if it weren't for the heavy automatic gate in front. Next came a huge gymnasium of a building with a broadly curving roof and playful porthole-like blue vents in the walls. It belonged to the system known as FLAG and served as the hinge of two cables that—like Tata's—stretched west to New York and east all the way to Japan. The locals called the place "Skewjack," after the surfers' camping ground that used to be on the site. Paling then led me into a narrow lane towered over by tall hedgerows. We had to squeeze to the left to let a tractor, loaded with hay, pass by. At a bend in the lane was a tan building with corrugated concrete walls, a hideous brutalist bunker. A NO TRESPASSING sign indicated that this station belonged to BT. Later I learned that it was designed according to off-the-shelf Cold War–era plans that presumed it would be underground. But when the Cornish granite proved too tough, BT put it aboveground instead. It looked ready for a war—the most menacing of the group.

At the crest of a hill I finally caught a glimpse over the hedgerows and saw pasture in all directions, dotted with an unlikely skyline of satellite dishes, mostly backup communication for the landing stations. We passed through a small suburban hamlet, and then the lane widened into a yard. A farmer in tall rubber boots had just pulled a red Land Rover out of a garage filled with tractors. His border collie raised his tail at me. On a wooden fence was a washed-out white sign with black letters that said WHITESANDS CABLE STATION. I followed Paling up the long driveway, with a potato field on one side and more pasture on the other. Dairy cows stuck their heads over the hedges, as if in the stocks. The farmer next door had a fire roaring in a steel drum, mixing the smell of peat smoke in with the manure. We thumped over a cattle grate and into the landing station's small parking lot. It had the shape of a house but was overscaled, like in a Texas suburb. Its exterior walls were faced in rough-hewn blocks of granite—at the request of the county planning commission— and there were green steel shutters. Beneath the eaves was a glass plaque that said ATLANTIC CROSSING. 1998. A GLOBAL CROSSING PROJECT.

Inside, rain slickers hung beside the doorway. The place smelled not unpleasantly of wet dog. Tia, a hefty spaniel, reclined in the corner. With its mismatched furniture, lime-green walls, maroon carpets, and a dropped ceiling, it had the feel of a technical shop rather than a slick high-tech command center. Giveaway maps from cable manufacturers were tacked to the wall. An old Global Crossing poster said "One Planet. One Network."

There was a cramped lobby, and a few private offices overlooking an idyllic Cornish scene of cows and emerald earth. The sound of a football game emanated from a television in the kitchen.

Paling had grown up in the area and had been with Global Crossing since 2000. Pushing forty, he was a big guy, more than six feet, with small blue eyes and a quiet face. He wore jeans, a stylish cardigan, and black skater shoes. If the Internet exchange guys tended toward the nerdy, most at home behind their screens, the undersea cable people were more likely to be the type who wouldn't hesitate walking into a sailors' bar in a foreign port. And indeed, Paling started out with BT in London, then spent time at sea laying and repairing cables, before returning to Cornwall to raise a family. His father had been "F1" with Cable & Wireless—the highest designation for a foreign officer—and had trained in Porthcurno. As a boy, Paling moved with his family among foreign stations, from Bermuda to Bahrain, the Gambia to Nigeria.

At Global Crossing, Paling wasn't only in charge of this station but of the field engineering for the entire undersea network, which included the link across the Atlantic, as well as major cables that connected the United States with South America, looping down both the Atlantic and Pacific coasts. Paling's eyes were bloodshot from a late night supervising, via conference call, equipment repairs on the link between Tijuana, Mexico, and Esterillos, Costa Rica. He knew the guys on the other end of the line well. His closest colleagues were on the other side of the world—which was also often the other end of the cable. This was typical. A cable across the ocean works like a single machine, with the equipment on one coast intricately linked with the equipment on the other. In the old days each cable would have an "order wire," a telephone headset labeled with the name of the city on the other end, providing a direct communications link. Today, the order wire has mostly been subsumed into the usual corporation communication system, although on an earlier visit I'd made to a cable station near Halifax, in Canada, I saw its progenitor in action. When I arrived in the morning a few minutes before the station manager, his colleagues on the other end of the cable—in Ireland—answered the doorbell and opened the gate remotely. Their systems were linked.

Ushering me into his office, Paling tossed his keys on the desk, beside a remote control yellow submarine, the size of a football. "For repairs," he said, nodding at it. Not really—it was his son's toy. We walked down the hall and into a room with wires strung overhead, racks of equipment arranged in narrow aisles, and the familiar roar of hot computer exhaust and blowing air conditioners. Paling led me straight to the far corner. A black cable came out of the floor and was attached with steel clamps to a heavy-duty frame set a few inches from the wall. It had been manufactured in New Hampshire. Over the course of a long passage through a series of machines worthy of Rube Goldberg, eight individual strands of fiber were woven with layers of rubber, plastic, copper, and steel. The cable was then spooled into steel trays the size of merry-go-rounds, like something stolen from Richard Serra's storehouse. A ship tied up to the factory's pier on the Piscataqua River, and the entire multithousand-mile length of cable was fed through a quarter-mile-long catwalk down to the water, into three cylindrical tanks in the hull. Out at sea, the ship paid the cable out its stern along a precisely planned path from a beach in

Long Island across the ocean to the sweeping arc of Whitesand Bay, a mile or so from here. Then it ran beneath the cows to the side of this large house, crossed the foundation, and popped up here. In its last foot, it had a label: AC-1 CABLE TO USA. For Paling, this was the plaque near his desk. For me, it was among the most amazing directional signs I'd ever seen. It pointed the way home, along a path that was physically utterly inaccessible—but that I'd in a way followed thousands of times before. "That's the cable going to the US," Paling said. The physical Internet couldn't get much more literal.

Having followed it across the ocean, I followed it a little farther, across the station. Paling showed me the PFE, or "Power Feed Equipment," a white box the size of a refrigerator that sent four thousand volts through the cable's copper shielding, to power the undersea repeaters that amplified the light signals. The companion machine on the other end of the cable, in Long Island, was adjusted to the same voltage, so that the flows of electrons met in the middle of the ocean and used the earth itself as a ground. "We're negative current, they're positive current," Paling said. It was a one-way flow of power, a simultaneous pushing and pulling.

The light through the cable was emitted (and received) from another bank of refrigerator-like machines, lined up in a row nearby. Paling found a spare length of yellow optical cable and plugged it into the "monitor" port of one of the machines, harmlessly tapping into the inbound light signal of one of the fibers. Then he plugged the other end of the cable into an optical spectrum analyzer—a desktop machine that looked like a Betamax, with a screen showing the waveforms of light, like on an EKG machine. "I like to think of it like a big jelly," he said about what was on the screen. "If you push that bit down"—he pointed to one of the waves—"then these all will go up. It's very much a case of playing around and trying to get that piece of jelly so each of the waves is at their best power." The technology was known as "dense wavelength division multiplexing." It allowed many wavelengths, or colors, of light to pass simultaneously through a single fiber. Each strand of fiber can be "filled up" with dozens of waves—each of which carries ten, twenty, or even forty gigabits per second of data. One of Paling's jobs was to tune the lasers to fit in more wavelengths, like a harmonizing chord, getting each one right so they all work well together.

Theoretically this can be done from anywhere, but Paling liked to be beside the machine, seeing the light with the analyzer. Making the process occasionally more difficult, any movement of the cable at the bottom of the ocean can change the way the waves move through the fiber, potentially knocking the whole arrangement out of whack, like static on an old TV. Once Paling had everything adjusted, he'd put the cable on a "confidence trial," generating artificial traffic to send through the fiber, and then looping that traffic, "backwards and forwards from here to America thirty times, or whatever." Things moved fast. The day I was there, one of the fiber pairs was "decommissioned" in preparation for an upgrade. New equipment was going to squeeze more twenty-gigabit waves into it, increasing the capacity of the whole cable.

"So the fiber's actually dark?" I asked.

"It's not dark, no," Paling said. "We call it 'dim.' There is power on those amplifiers. They are putting out ASE"—amplified spontaneous emission. "Noise. If you put a meter on there, you'd see light. But there's no band noise. It's only just background noise." A flickering.

As Paling explained all this, he absentmindedly flipped open a protective plastic shield and tapped his finger against one of the "lit" fibers. All across Europe—if not the whole Eastern Hemisphere—there were millions and millions of strands of fiber. They merged and merged, again and again, emerging out of Telehouse in a thick bundle, then heading here. The final merge could be read in the yellow cables plugged into the front of this machine: many fibers went in, but ultimately only four came out. It was these four that would cross the ocean. They were the thickest veins at the end of a continent of capillaries—in terms of what they contained, but certainly not their physical size. It was just before noon. The European markets were open, but New York was still waking up. Paling's lips were moving, but all I could concentrate on was his finger tapping on the cable. Short of renting a submarine, this was as close as I was going to get to a physical transatlantic link.

Our final stop was back across the hall. We'd followed the cable from where it came out of the dirt to the major subsea equipment. Now we were looking at the "backhaul," the links from the station to the rest of England. One rack was labeled SLOUGH, a bland London suburb, not far from Heathrow, where Equinix had its biggest UK data center (and the original version of *The Office* is set). The one next to it was labeled DOCKLANDS. Whatever the size of the world, not for the first time, I thought to myself how small the Internet itself could seem.

Later that afternoon, after Paling had gotten back to work and we'd said our good-byes, I drove out to Land's End. There's a theme park there with a fake medieval street, but it was past the season and almost everything was closed up, except for a famous photo concession near the edge of the cliff, looking out over the ocean. For fifteen quid, you chose letters to spell out the name of your hometown and slid them into one of those signs that point to far-off places and list their distances. The photographer in a thick wool sweater snapped your picture, and a few weeks later the print arrived in the mail. Two of the destinations on the sign were permanent: John O'Groats, the most northerly place on mainland Britain (874 miles), and New York (3,147 miles). I figured since New York was already there I could ask for a different place and thought, what the hell: Would he mind if I put "The Internet" up there, and call it two miles away? For fifteen quid he said he didn't mind at all, and he knew exactly why I'd asked. He knew the cables well. He'd watched the ships come by the water below. After my official shot, he offered to take an extra one with my phone.

Back in the warmth of the car, I emailed the picture to a few people in New York. I couldn't help but think what that meant: the connection with the nearby cell tower, the backhaul to the Docklands, the U-turn to Cornwall, the quick pass through the cable landing station, the long journey to Long Island, into 60 Hudson, and then to my own email server in Lower Manhattan, before splattering out to its recipients. I knew these physical paths existed. But I also knew the Internet was still wily, diverse, multitudinous. I couldn't say which path that photo went; it could just as easily have gone through the big Tata cable, which lands farther up the coast. The movement of a single clump of data was difficult to nail down, but that didn't make the particularities along its path any less real. I was struck again by this challenge of catching lighting in a bottle—of nailing the Internet down, if only for a minute. There was still this gap between the physical and the virtual, the abstract of information and the damp breeze off the sea.

It took me a few months, but back in New York I eventually found a free afternoon to drive out to the beach and look for the other end of Atlantic Crossing-1. I decided not to call ahead. Paling had been a great host in Cornwall, and it felt like over-kill to request to see yet another landing station—much less the matched pair. The town of Shirley on Long Island was AC-1's officially listed landing point, but that left a fairly wide swath of beach where it could actually be. My wife and daughter came along, and they humored me as I nosed around at the edge of the parking lot of a public beach, kicking sand around like a scavenger. I eventually found a weather-beaten plastic post with a warning about a buried fiber-optic cable—but I couldn't be sure if it was my cable. As we headed back toward the city, me a little disappointed that the landscape hadn't been so easy to read, a building a mile or so from the beach caught my attention out of the corner of my eye. It sat on the edge of a suburban subdivision and looked basically like a house except for being too big, and it had telltale steel vents beneath its eaves. I swung around at the next light and pulled up in front. It had a sturdy gate, big surveil-lance cameras, and a few cars in a small parking lot—including a white pickup truck with a tool case in the back and an AT&T logo on the door. That's when I noticed the mailbox: hardware store stickers spelled "TT," with the faint outline of the ripped-off "A" still visible. This wasn't my cable, but it was my kind of place—clear enough to particularize the single pathway, to layer the fluid geography of the Internet on the sandy ground of Long Island, and across the ocean in between.

———————

In the interval, I had been holding out for word from Simon Cooper at Tata, about a new cable landing on a beach, somewhere. The email from the press person in Mumbai came on a Thursday morning and said that, depending on the weather, the landing was planned for that coming Monday. Somewhere near Lisbon. Where exactly she wasn't sure. I didn't reply right away. Instead, I started looking for a plane ticket.

That Sunday morning I arrived in Portugal, crossed the Tagus River from Lisbon and turned west, again toward the Atlantic. I followed the sandy Costa da Caparica south for a few miles before heading back inland, into a neighborhood of modest weekend villas. Tata's cable landing station sat slightly back from the road behind a high security fence. Unmarked and slightly sinister looking, it had thick concrete walls and windows with heavy steel frames. It could be mistaken for the home of an arms dealer, or perhaps the listening post of a top-secret intelligence service. It was far bigger than Paling's station in Cornwall—a textbook example of Tyco's excess. I pressed the intercom button and waited as the gate swept open, then mustered the extent of my jet-lagged concentration to ease the clutch on the manual transmission and slide into the small parking lot, crowded on a Sunday morning.

Rui Carrilho, the station manager, was a compact guy in his

early forties. He wore a bright blue polo shirt, jeans, and leather oxfords, as if dressed for a Sunday stroll with his wife. He was not happy to see me. I was there on the invitation of his boss, Simon Cooper, but this was not a good week for visitors. Despite the calm winds and clear sky, he was in the middle of a shit storm. There was the reason I had come: the arrival on the beach of the West Africa Cable System, or WACS, which would soon reach from this hillside down the coast of Africa. But the station was also hosting a pair of technicians from Tyco headquarters in the United States, who had been working around the clock on the final commissioning of WACS's direct competitor, the Main One cable, which followed nearly the identical route. They had been on call around the clock, waiting for direction from Tyco's cable-laying ship, the *Resolute,* bobbing somewhere off the coast of Nigeria, the cable hauled up into its workshop, as its own technicians struggled to get out the kinks. And on top of all that, Tata brass had been pushing Carrilho to complete upgrades on the station's third cable, which ran beneath the Bay of Biscay to England, crossed AC-1 somewhere in the deep, and landed at another large former Tyco station near Bristol. The cable station staff had been sleeping on the office floor all week and eating their evening meals at a restaurant nearby—sometimes joined by the exhausted Tyco engineers from New Jersey. Carrilho sat at the head, leading his men like the air force officer he once was. But the bags under Carrilho's eyes—and the nervous intensity with which he gripped his BlackBerry and his Camels—made it clear: there was a lot going on. And I, a tourist, had walked into the middle of it.

I'd hardly stepped into the place when he turned me back out the door and into the station minivan. "I am going to show you where the beach landing is, so you can get there on your own," he said, eager to get rid of me as soon as possible. We headed toward the ocean, following a leafy boulevard, beneath which ran the cable from the beach (and far beyond). At the bottom of a steep hill was a tiny beach hamlet, a dusty turnaround where mangy dogs slept in the sun. Carrilho put on a hard hat and orange safety vest and mounted an orange flashing light to the top of the van. A pair of old men in plaid shirts shifted their gaze from the sea to us. A square of sand the size of a beach blanket had been scraped away to reveal a manhole, opening into a concrete vault. The manhole cover was stamped "Tyco Communications." The vault had been dug a decade earlier in preparation for a cable that never arrived and had sat empty since. A red tent had been erected beside it, to house a temporary workshop.

The next day, the cable-laying ship *Peter Faber*—specially designed for "near-shore operations"—would steam over from Lisbon with two miles of cable in its hold. It would be dragged up on the beach by a diver and affixed to a heavy steel plate inside the manhole. The *Peter Faber* would then head out to sea a couple of miles, turn slightly to the south, and drop the loose end over the side. A couple months later a much larger ship would come back to pick it up with a grappling hook, fuse it to the end of the remaining nine thousand miles of cable it carried in its hull tanks, and turn southward, following a precisely prescribed route above underwater canyons and along the edges of invisible cliffs. For the people of South Africa, Namibia, Angola, the Democratic Republic of Congo, the Republic of Congo, Cameroon, Nigeria, Togo, Ghana, Côte d'Ivoire, Cape Verde, and the

Canary Islands—the cable's successive landing points—this singular spot on the earth would soon affix this continent to the other. That, at least, was the plan for the next days and months. The plan for the next hour was lunch.

Almost on top of the manhole was a beachside restaurant with Coca-Cola umbrellas on the patio. At a long table inside, the underwater construction team had assembled. With their red jumpsuits, sea-weathered faces, and windblown hair, they looked like a band of pirates. I helped myself to a seat beside one wearing a bandanna over his unruly black hair and a gold loop in his ear. Carrilho sat at the other end of the table, between the wizened construction manager, Luis, with a yellow mustache, and his foreman, Antonio, who looked a bit like Tom Cruise and had the determination and emotional intensity of a preschooler. They sketched the next day's landing plans on the white paper tablecloth until an enormous pot of fish stew arrived, along with glasses of Super Bock, the Portuguese lager. The conversation had been a mix of Portuguese and Spanish and had stopped for the soccer game on the TV. But when it came time to toast the success of the operation, they used the English term: to the "beach landing!"

Landing day dawned cold and bright, the blues of the ocean and sky in apparent competition to be the deepest. Carrilho had his hard hat and vest on, and he'd brought along one of the young guys from the station, who wore a big camera around his neck. He paced in and out of the café, ordering espressos and checking the horizon. The construction team arrived by boat from a port a few miles down the coast, bouncing in on their inflatable skiff like a platoon of marines. A group of Angolan day laborers had assembled, and Luis handed out red polo shirts from a big cardboard box. A pair of British engineers, in fleece and cargo pants, kept to themselves, perched on the edge of a small sand cliff. They worked for Alcatel-Lucent, the telecommunications conglomerate that manufactured the cable and owned the ships that would lay it on the ocean floor.

A large Hyundai excavator was parked down by the water, its articulated arm raised in a curled salute, a sign reading CARLOS propped against its windshield. Carlos himself sat inside the cab, leaning forward against the dashboard. Normally he demolished historic buildings in Lisbon—delicate work. Luis had worked with him before. "He can scratch your nose with his bucket, and you wouldn't mind," Luis said, wiggling a finger at me. The previous day, Carlos had dug a deep trench in the beach, leaving a sandcastle as big as his machine. In its depths was the mouth of a steel conduit that ran back up to the manhole; the fiber-optic cable would be pulled through it like a string through a straw.

Just before nine one of the divers hopped out of the skiff and into the surf. Under his arm, he carried a length of lightweight green nylon line. He high-stepped through the waves up onto the beach and handed the line to one of the laborers. There was no handshake or ceremony to mark this first moment of physical connection, the initial link between land and sea that would be leveraged into a nine-thousand-mile path of light—and, its backers hoped, a stream of information that would transform a continent. Carrilho stopped his pacing on the café patio to watch. Soon after, the blue hull of the cable ship *Peter Faber* steamed into view from the north, its large white antennae dome perched like a Ping-Pong ball at the peak of its superstructure. Longer than a tugboat and sleeker than a trawler, its GPS-controlled propulsion system allowed it to hover in place, even in rough conditions. It parked almost a kilometer offshore, precisely lined up with the beach manhole, and wouldn't move from that spot for a day and a half.

The skiff headed out to meet it, paying out the lightweight green messenger line as it went. Two dogs frisked on the beach, jumping back and forth across the thick rope. A bulldozer chugged down to the water, and the rope was knotted to its hitch. It began a series of slow processions down the beach, parallel to the water, dragging the line around a pulley, a hundred meters at a time in from the ship. The dozer would chug out at a walking pace, drop the knot, and then reverse along the same tracks, to pick up the next length. The fiber-optic cable itself soon began coming off the ship, suspended just below the surface of the water by a necklace of orange buoys—the modern-day version of the "casks" used in Porthcurno, in 1919. As each buoy reached the shore, a laborer skipped into the surf and untied it from the cable.

Carrilho and I watched the action from the restaurant patio, sitting at separate tables. He had a tab open, and I joined him in a steady alternating rhythm of espresso and beer. A soft onshore breeze brought the pleasantly nautical smell of the skiff's two-stroke engines. It had been working hard keeping fishing boats from crossing the cable, patrolling back and forth like a border collie. By lunchtime, the bulldozer had completed its slow laps, and the cable arced in from the ship beneath its necklace of orange buoys. Wearing thick gloves, the laborers manhandled it into the mouth of the conduit, straining under its weight. They laid it out in an S-pattern down to the surf, in case the ocean wanted a little more for itself. I emailed Simon Cooper a picture of the action and captioned it "taken forty-five seconds ago." I got a message back a few minutes later: "And viewed on my BlackBerry whilst roaming in Tokyo."

With the cable in position, the diver walked back into the surf holding a knife. Bobbing his head under the waves, he began to cut the orange buoys free, so that the cable could drop to the seafloor. With each slice, a buoy popped a few feet into the air, and then shot south with the breeze. By the time he was a hundred or so meters out, I could no longer see him, only his handiwork: orange buoys popping up out of the surf like beach balls, the skiff chasing each one down. When he arrived at the cable ship, the Dutch crew gave him a few cookies and a glass of juice, and then he jumped back into the ocean and swam the kilometer back to shore. Back on the beach, chest heaving and eyes wide, he lit a cigarette.

I walked around to the side door of the restaurant, where the two English engineers were hard at work on the cable. They'd driven from Alcatel-Lucent's offices in London, in a station wagon full of tools. Matt was tall, with a square head, a big potbelly, and a jolly voice. He lived in Greenwich—"home of time," he sang—and was eager to get back there for his son's birthday that weekend. Mark was rougher, with a gold tooth and a big tattoo on an arm Popeye would envy. He'd spent his working life all over the world on Alcatel's behalf, in places like Bermuda ("perfect"), California ("lovely"), Singapore ("a great city, if you like sitting down in the evening with a beer"). In blue Alcatel polo shirts and cargo work pants, the two of them went at the cable with hacksaws. It was armored with two layers of woven steel mesh that had to be stripped off before they made the "joint" in the manhole. They put their full weight into peeling back the casing, as if butchering a shark. While they worked, a fisherman in a flannel shirt and rubber boots knocked on the kitchen door. He had a stiff tote bag filled with two glistening dourada, or sea bream, the same as in yesterday's fish stew. The chef took them. Matt yelled into his telephone: "We've got twenty-five in the manhole and another twenty on the beach," meaning meters of cable—enough slack underground to finish the joint, and enough on the beach to allow the cable some give in a storm.

Once the cable was stripped back to its pinkish intestines, Matt and Mark fed it into the red tent to begin the work of fusing the fibers together. Matt set a cup of tea beside him on the workbench and began using a tool that looked like a corkscrew to shear off the cable's plastic inner core, which surrounded a perfect tube of shiny copper. Inside of that was another layer of black threads; inside of each, colored rubber; inside of the rubber, the fiber itself. With the removal of each successive layer, the work became increasingly delicate: he worked first like a butcher, then a fisherman, then a sous-chef, now finally a jeweler, as he held each fiber between his pursed lips. When the fiber itself was finally visible, the eight strands glinted in the sunlight, each a hundred and twenty-five microns wide. He put baby powder in his palm and ran the end of each fiber through it, like a violin's bow, to clean off any residue.

Then he began to fuse each to its shoreside mate. There were eight strands, each a different color, looped around the work-table. One at a time, Matt placed a strand inside a machine that looked like a hole puncher. A small screen magnified the fibers' alignment, and he adjusted it so the two ends lined up, like the hand of God in Michelangelo's fresco. Then he pressed a button that baked the two ends together, sipping tea with his pinky up while the machine did its work. Then he slid a little protective plastic sheath over the now-continuous strand and mounted it in a dainty rack, like a fishing tackle. With today's technology, each fiber could transmit more than a terabit of data per second, on an undersea journey of two-tenths of a second.

I had been watching from just outside the red tent that shel-tered the makeshift workshop, and Carrilho came up beside me, intently watching Matt's delicate work. "That's the fiber," Car-rilho said. "That's what makes the money."

With all eight strands finally spliced together, Matt mounted them inside a finely machined black steel case, with two large red laser warning stickers, and an elegant Alcatel-Lucent plaque, with *Origin France* written in energetic italic script—the hood ornament of Tata's $600 million wire. Mark had been working in the manhole, doing the heavy labor of tightening the steel mesh cable around a heavy steel plate built into its wall. Matt passed the case down to him, to be mounted inside.

A car pulled up behind me and Carrilho, and a man in a pressed white shirt and tie, on his way home from work, stepped out. He looked into the manhole, at the equipment arrayed in-side the tent, and the ship steady offshore.

"A cable? To Brazil?" he asked.

"Africa," Carrilho replied.

The commuter raised his eyebrows, shook his head, and went home to dinner. For the people of this seaside village, this was a temporary disruption, a few days of bulldozers on the beach and some extra trucks in the municipal lot. By the end of the week, the manhole would be covered, and the cable to Africa would be forgotten under the sand.

7

Where Data Sleeps

The Dalles, Oregon, has always been a special kind of cross-roads, a place where geography has repeatedly forced the hand of infrastructure. Its odd name—it rhymes with neither "balls" nor "bells"—comes from the French word for "flagstone" and refers to the rocks in the mighty Columbia River, which narrows here before plunging through the great gap in the Cascade Range known as the Columbia River Gorge. Everything here has followed from that.

When Lewis and Clark arrived in 1805 on their exploration of the west, they found the largest Native American gathering place in the region. During the annual salmon runs, the population swelled to nearly ten thousand, about the same size as the town today. For a while the Oregon Trail ended in The Dalles, where western settlers faced the uncomfortable choice of mule-packing around 11,249-foot-tall Mt. Hood, or braving the Columbia's rapids. The Dalles was the choke point in the path of western migration. It still is.

From my motel room the landscape looked like a battlefield between geology and industry. In the backdrop were the lumpy tan foothills of the Cascades, covered in wisps of fog on a rainy late-winter day. In the near distance was the Union Pacific rail yard, where freight trains stopped and stuttered before descending through the modern alignment that hugs the basalt cliffs of the gorge. Parallel to the tracks is Interstate 84, the first major east/west route across the mountains until Interstate 80 in California, nearly six hundred miles to the south. Truck traffic rushed by all night, heading west to Portland, or east toward Spokane, Boise, and Salt Lake. The river itself was wide and gurgly. I watched cars streaming antlike across the Dalles Bridge, just downstream from the Dalles Dam, a small piece of the vast hydroelectric system built by the Army Corps of Engineers and marketed by the Bonneville Power Administration, whose high-tension lines lace the hills. The Dalles is a key node in the power grid of the entire western United States. Most notably, it is the starting point for a 3,100-megawatt transmission path, known as the Pacific DC Intertie, which transfers hydroelectric power from the Columbia River basin to Southern California, like a massive extension cord dragged up from Los Angeles. Its plug is the Celilo Converter Station, just over the hill from my motel room. The Dalles might be a small place but you wouldn't say it's off the beaten path. It *is* the beaten path: an infrastructural confluence, where the inescapable topography of the Cascades and the Columbia River have forced together salmon, settlers, railroads, highways, power lines, and, as it turns out, the Internet.

I had come to The Dalles because it is home to one of the Internet's most important repositories of data, as well as being the de facto capital of a whole region devoted to storing our on-line selves. The place struck me as a sort of Kathmandu for data centers, a foggy town at the base of a mountain that happened to be the perfect jumping-off point for an exploration of the mas-sive buildings where our data is stored. Even better, The Dalles was mysterious and evocative enough—a natural nexus—to highlight these buildings' strange powers. A data center doesn't merely contain the hard drives that contain our data. Our data has become the mirror of our identities, the physical embodi-ment of our most personal facts and feelings. A data center is the storehouse of the digital soul. I liked the idea of data centers tucked away up in the mountains like wizards—or perhaps war-heads. And Kathmandu felt right in another way: I was looking for enlightenment, for a new sense of my digital self.

Up until now I'd focused mainly on exchange points: the In-ternet's central hubs, the places where networks meet to become an Internetwork. My mind had filled with accumulated images of corrugated steel buildings, yellow fiber-optic cables, and base-ment vaults. But data centers presented a different challenge on this journey. They seemed to be everywhere. As I considered this, a more schematic image of the Internet came to mind, of two funnels fused at their narrow ends, like a siamese vuvuzela. The exchanges sit at the narrow spot in the middle. There aren't very many of them, but they are the choke points for the vast majority of traffic. One funnel pulls in all of us: the billions of "eyeballs" scattered around the world. The other funnel catches the buildings where our data is stored, processed, and served. Data centers are what's on the other end of the tubes. They're only able to exist in faraway places thanks to the thicket of networks everywhere else.

It used to be that we kept our data on our (actual) desks, but as we've increasingly given up that local control to far-off professionals, the "hard drive"—that most tangible of descriptors—has transformed into a "cloud," the catchall term for any data or service kept out there, somewhere on the Internet. Needless to say, there is nothing cloudlike about it. According to a 2010 Greenpeace report, 2 percent of the world's electricity usage can now be traced to data centers, and that usage is growing at a rate of 12 percent a year. By today's standards, a very large data center might be a 500,000-square-foot building demanding fifty megawatts of power, which is about how much it takes to light a small city. But the largest data center "campus" might contain four of those buildings, totaling more than a million square feet—twice the size of the Javits Center in New York, the same as ten Walmarts. We've only just begun building data centers, and already their accumulated impact is enormous.

I know this intuitively, because a lot of this data is mine. I have gigabytes of email storage in a data center in Lower Manhattan (and growing every day); another sixty gigabytes of online backup storage in Virginia; the cumulative traces of countless Google searches; a season's worth of episodes of *Top Chef* downloaded from Apple; dozens of movies streamed from Netflix; pictures on Facebook; more than a thousand tweets and a couple hundred blog posts. Multiply that around the world and the numbers defy belief. In 2011, Facebook reported that nearly six billion photos were uploaded to the service every *month*. Google confirms at least one billion searches per day—with some estimates tripling that number. All that has to be processed and stored somewhere. So where does it all go?

I was less interested in the aggregate statistics than in the specifics, the parts of all this online detritus that I could touch. I knew that data centers which once occupied closets had expanded to fill whole floors of buildings; that floors had grown into subdivided warehouses; and that warehouses have transformed into purpose-built campuses, as in The Dalles. What had before been afterthoughts, physically speaking, had now acquired their own architecture; soon, they'd need urban planning. A data center was once like a closet, but now was more like a village. The ever-increasing size of my own appetite for the Internet made it clear why. What was less clear to me was where. What were these enormous buildings doing way up on the Columbia Plateau?

The Internet's efficiency at moving traffic—and the success of exchange points at serving as hubs for that traffic—has left the question of where data sleeps remarkably open-ended. When we request information over the Internet, it has to come from somewhere: either another person or from the place where it's being stored. But the everyday miracle of the Internet allows all that data to, in theory, be stored anywhere—and still the stuff will find its way back to us. Accordingly, for smaller data centers, convenience rules: they are often close to their founders or their customers, or whoever finds a need to visit them to tweak the machines. But as it happens, the bigger a data center gets, the thornier the question of location becomes. Ironically for such massive, factory-like buildings, data centers can seem quite loosely connected to the earth. But still they cluster.

Dozens of considerations go into locating a data center, but they almost all come down to making it as cheap as possible to keep a hard drive—much less 150,000 of them—spinning and cool. The engineering of the building itself, especially how its temperature is controlled, has a huge impact on its efficiency. Data center engineers compete to design buildings with the lowest "power usage effectiveness," or "PUE," which is sort of like the gas mileage in a car. But among the most important external variables in a good PUE is a building's location. Just as a car will get better gas mileage in a flat, empty place compared with a hilly city, a data center will run more efficiently where it can draw in outside air to cool its spinning hard drives and powerful computers. But because data centers *can* be anywhere, seemingly small differences become amplified.

Siting a data center is like the acupuncture of the physical Internet, with places carefully chosen with pinpoint precision to exploit one characteristic or another. As competitive companies thrust and parry for advantage, it becomes clear that some places are better than others, and the result is geographic clusters. The largest data centers begin piling up in the same corners of the earth, like snowdrifts.

Michael Manos has built more data centers than perhaps anyone—by his count around a hundred, first for Microsoft and later for Digital Realty Trust, a major wholesale developer. He is a big, fair-skinned, good-humored guy and he talks a mile a minute, like John Candy playing a commercial real estate agent.

That suits the data center game, which is about finding a deal and driving your stake. When he joined Microsoft in 2005, the company had about ten thousand servers spread out in three separate facilities around the world, running their online services like Hotmail, MSN, and Xbox games. By the time Manos left four years later, he had helped expand Microsoft's footprint to "hundreds of thousands" of servers spread around the world in "tens" of facilities—"But I still can't tell you how many," he told me. The number was still a secret. It was an expansion of unprecedented scale in Microsoft's history, and one that to this day has been matched only by a handful of other companies. "Not a lot of people on the planet are dealing with these size and scale issues," Manos said. Even fewer have scoured the world as Manos has.

At Microsoft, he built a mapping tool that considered fifty-six different criteria to generate a "heat map" indicating the best location for a data center, shaded from green (for good) to red (for bad). But the trick was getting the scale right. At the state level, a place like Oregon looked horrible—mainly because of environmental risks, like earthquakes. But when he zoomed in, the story changed: the earthquake zone is on the western side of the state, while central Oregon has the benefit of being cold and dry—perfect for cooling hard drives using outside air. Surprisingly, what got almost no weight in the equation was the cost of the land itself, or even the cost of the actual building. "If you look at the numbers, eighty-five-ish percent of your cost is in the mechanical and electrical systems inside the building," Manos explained. "Roughly seven percent, on average, is land, concrete, and steel. That's nothing! People always ask me, 'Is it better to build small and tall or big and wide?' It doesn't matter. At the end of the day, real estate and the biggest construction costs are literally not an issue for most of these buildings. All your cost is in how much gear can you stick in your box." And then, of course, how much it costs to plug it in—what data center people call "op-ex," the operating expenses. "A data center guy is always looking for two things," Manos said. "My wife used to think I was always looking at the scenery, but actually I was looking at the power lines, and for fiber hanging from those power lines." In other words, he was looking for the view outside my window in The Dalles.

Beginning in the late 1990s, the Bonneville Power Administration had begun mounting fiber-optic cables along its long-haul transmission lines, an amazing network that crisscrossed the Northwest and came together in The Dalles. It was a tricky job, often requiring helicopters to string cable on high towers in rough country, and while the power company's leaders' primary goal was improving internal communications, they saw it was only incrementally more expensive to install extra fiber—far more, in fact, than they needed for the company's own use. To the strenuous objection of the telecommunications companies, who didn't believe a government-subsidized utility should be competing with them, the BPA soon began leasing that extra fiber out. It was a big, robust communications system, a regional sweep of heavy-duty fiber protected from errant backhoes on its perch high up on the power lines—catnip for data center developers.

Microsoft tapped into it from a town called Quincy, up the road from The Dalles in Washington State. "It was the greenest spot in the United States for us," Manos said, referring to his heat map, rather than trees or environmental considerations. Like The Dalles, Quincy was near the Columbia River and nestled in the tangle of the Bonneville Power Administration's power and fiber infrastructure. Not surprisingly, Microsoft wasn't alone for long. Soon after breaking ground on its 470,000-square-foot, 48-megawatt data center (since joined by a second building), what Manos calls the "Burger King people" showed up—the second movers, the companies who wait until the market leader has built in a particular location, and then build next to them. In Quincy, these included Yahoo!, Ask.com, and Sabey, a wholesale data center owner. "Within eighteen months, you had this massive, almost three billion dollars' worth of data center construction going on in a town that was predominantly known for growing spearmint, beans, and potatoes," Manos said. "When you drive through downtown now, it's just big, giant, open farm fields and then these massive monuments of the Internet age sticking out of these corn rows." Meanwhile, down the road in The Dalles, one of Microsoft's biggest competitors was writing its own story.

The Dalles had been a crossroads for centuries, but around 2000, at the crest of the broadband boom, it seemed as if the Internet was passing it by. The Dalles was without high-speed access for businesses and homes, despite the big nationwide backbones that tore right through along the railroad tracks, and the BPA's big network. Worse, Sprint, the local carrier, said the city wouldn't get access for another five to ten years. "It was like being a town that sits next to the freeway but has no off-ramp," was how Nolan Young, the city manager, explained it to me in his worn office, grand but fluorescent lit like a high school principal's, inside the turn-of-the-century Dalles City Hall. Wizened and soft-spoken, with a hobbitlike pitch to his voice, Young had shrugged at the sight of my tape recorder. Like any veteran politician, he was used to nosy journalists—although more than a small town's share had been through here recently.

The Dalles had felt the brunt of the industrial collapse of the Pacific Northwest, and the Internet's neglect added insult to injury. "We said, 'That's not quick enough for us! We'll do it ourselves,'" Young recalled. It was an act of both faith and desperation—the ultimate "if you build it they will come" move. In 2002, the Quality Life Broadband Network, or "Q-Life," was chartered as an independent utility, with local hospitals and schools as its first customers. Construction began on a seventeen-mile fiber loop around The Dalles, from city hall to a hub at the BPA's Big Eddy substation, on the outskirts of town. Its total cost was $1.8 million, funded half with federal and state grants, and half with a loan. No city funds were used.

The Dalles's predicament was typical of towns on the wrong side of the "digital divide," as politicians call poorer communities' lack of access to broadband. The big nationwide backbones were quickly and robustly built, but they often passed through rural areas without stopping. The reasons were both economic and technological. Long-distance fiber-optic networks are built in fifty-odd-mile segments, which is the distance light signals in fiber-optic cables can travel before needing to be broken down and reamplified. But even at those "regeneration" points, siphoning off the long-distance signals for local distribution requires expensive equipment and a lot of person-hours to set up.

High-capacity, long-distance fiber-optic networks are therefore cheaper to build and to operate if they zoom straight through on their path between hubs. And even if they can be induced to stop, a small town doesn't have the density of customers needed to push it up the priority list of construction projects for a national company, like Sprint. A "middle mile" network bridges that gap, by laying fiber between a town and the nearest regional hub, connecting small local networks to the long-distance backbones. Network engineers call this the "backhaul," and there's no Internet without it. Q-Life was a textbook example of the middle mile—although in The Dalles, the middle mile was actually closer to four miles, from the center of town to the Big Eddy substation, where the BPA's fiber converged.

Once Q-Life's fiber was in place, local Internet service providers quickly swooped in to offer the services Sprint wouldn't. Six months later, Sprint itself even showed up—quite a lot sooner than its original five-year timeline. "We count that as one of our successes," Young said. "One could say they're our competitors, but now there were options." But the town couldn't have predicted what happened next. At the time, few could have. The Dalles was about to become home to the world's most famous data center.

In 2004, just a year after the Q-Life network was completed, a man named Chris Sacca, representing a company with the suspiciously generic name of "Design LLC," showed up in The Dalles looking for shovel-ready sites in "enterprise zones," where tax breaks and other incentives were offered to encourage businesses to locate there. He was young, sloppily dressed, and interested in such astronomical quantities of power that a nearby town had suspected him as a terrorist and called the Department of Homeland Security. The Dalles had a site for him, thirty acres next to a decommissioned aluminum smelter that itself once drew eighty-five megawatts of power—more than the everyday needs of a city many times its size.

As negotiations began, Sacca wanted total secrecy, and Young started signing nondisclosure agreements. The cost of the land itself wasn't much at issue (as Manos could have predicted). It was all about power and taxes. The local congressman was called in to help convince the Bonneville Power Administration to steepen its discounts. The governor had to approve the fifteen-year tax break Design LLC demanded, given the hundreds of millions of dollars of equipment it planned to install in The Dalles. But any reasonably sized community in Oregon might have come up with the power and the incentives. The ace in the hole that made Design LLC's heat map glow bright green over The Dalles was of the town's own making: Q-Life. "It was visionary—this little town with no tax revenues had figured out that if you want to transform an economy from manufacturing to information, you've got to pull fiber," Sacca later said. In early 2005, the deal was approved: $1.87 million for the land and an option for three more tracts. But still Young had to keep the secret, even after construction began. "I had signed so many agreements that there was a point when I was standing at the site, and someone said, 'I see they're building . . . rrrggrrr . . . there.' And I said, 'What, I don't see anything!'" But the secret's out now: Design LLC was Google.

It's become a cliché that data centers adhere to the same rules as the secret cage matches in the movie *Fight Club:* "The first rule of data centers is don't talk about data centers." This tendency toward the hush-hush often bleeds into people's expectations about the *other* types of the Internet's physical infrastructure, like exchange points—which are actually quite open. So why all the secrecy about data centers? A data center is a storehouse of information, the closest the Internet has to a physical vault. Exchange points are merely transient places, as Arnold Nipper pointed out in Frankfurt; information passes through (and fast!). But in data centers it's relatively static, and physically contained in equipment that needs to be protected, and which itself has enormous value. Yet more often the secrecy isn't because of concerns over privacy or theft, but competition. Knowing how big a data center is, how much power it uses, and precisely what's inside is the kind of proprietary information technology companies are eager to keep under wraps. (And indeed, Manos and Sacca very well might have run into each other, crisscrossing the Columbia River Valley in search of a site.) This is especially true for data centers built and owned by single companies, where the buildings themselves can be correlated to the products they offer. A culture of secrecy developed in the data center world, with companies fiercely protecting both the full scope of their operations, and the particularities of the machines housed inside. The details of a data center became like the formula for Coke, among the most important corporate secrets.

As a consequence, from a regular Internet user's perspective, where our data sleeps is often a difficult question to answer. Big web-based companies in particular seem to enjoy hiding within "the cloud." They are frequently cagey about where they keep your data, sometimes even pretending not to be entirely sure about it themselves. As one data center expert put it to me, "Sometimes the answer to the question 'where's my email?' is more quantum than Newtonian"—a geeky way of saying it appears to be in so many places at once that it's as if it's nowhere at all. Sometimes the location of our data is obscured further by what are known as "content delivery networks," which keep copies of frequently accessed data, like popular YouTube clips or TV shows, in many small servers closer to people's homes, just as a local store keeps popular items in stock. Being close minimizes the chances of congestion, while also bringing bandwidth costs down. But generally speaking, the cloud asks us to believe that our data is an abstraction, not a physical reality.

But that's disingenuous. While there are moments when our online life really has discombobulated, with our data broken into ever-smaller pieces to the point that it's theoretically impossible to know where it is, that's still the exception. It's a quarter truth that data center owners seize upon in a deliberate attempt at directing attention away from their actual places—whether for competitive reasons, because of environmental embarrassment, or for other notions of security. But what frustrates me is that feigned obscurity becomes a malignant advantage of the cloud, a condescending purr of "we'll take care of that for you" that in its plea for our ignorance reminds me of slaughterhouses. Our data is always *somewhere,* often in two places. Given that it's ours, I stick to the belief that we should know where it is, how it ended up there, and what it's like. It seems a basic tenet of today's

Internet: if we're entrusting so much of who we are to large companies, they should entrust us with a sense of where they're keeping it all, and what it looks like.

Nolan Young was happy to show me his data center, good public official that he is. Without thinking too much about it, soon after The Dalles's fiber loop was completed, Young carved out a little space in the basement of City Hall where customers could put a rack of equipment and connect to one another—like a mini-Ashburn on the Columbia. Of course I wanted to see it; it sounded like a nice little piece of the Internet. "It's just boxes and lights, but if you want!" Young said, and he fetched the key from his assistant. The Dalles's town court was across the hall from Young's office, and we walked past a sullen teenager waiting outside with his mother, then down the grand staircase at the center of the building, out the front steps, and around to a little side door into the basement. There was a small vestibule, linoleum tiled and fluorescent lit. Young opened a steel door, and I was greeted by the roar of blowing air and the old familiar electric smell of networking equipment. The Dalles's data center may have looked like a glorified closet, but it struck me as an exchange point in its purest incarnation: just a bunch of routers plugged into one another in the dark. Young cheerily pointed out the pathways: "The customers come in here, jump onto our fiber, connect to each other, do whatever else they do, then pop back on our fiber out to Big Eddy and go wherever they go! The technical end of it is beyond me. I just know all that stuff goes to one spot and then it shoots out." This was among the smallest rooms you might feign to call a data center—in the shadow of one of the largest. But in its homespun simplicity it was a vivid confirmation that the Internet is always somewhere.

Working with the press people at the Googleplex—"Design LLC's" headquarters in Silicon Valley—I had arranged a visit to their massive data center that afternoon, but Young warned me not to expect much. "I can pretty much guarantee that the closest you're going to get is the lunchroom," he said. We said our good-byes, and I made sure Young had my email address. "I do. Now we're connected! I'll jump on that fiber, and there we are."

From City Hall I drove the five minutes across town, over the interstate and into an industrial neighborhood along the banks of the Columbia, nearly at the entrance to the gorge. The vast campus was visible from a distance, sitting beside the highway. It looked like a prison, thanks to its towering security lights, loosely spaced beige buildings, and strong perimeter fence. Huge power lines hemmed the campus into the base of the mountains, whose middle reaches were still dusted with snow, and their tops obscured by fog. On the corner was an animal shelter. Across the street was a concrete plant. Every hundred yards or so I passed a white safety pylon with an orange top that said BURIED FIBER-OPTIC CABLE—Q-LIFE. I drove past a DEAD END sign and buzzed the intercom on the outside of a double gate. It opened, and I parked in front of a security building the size of a house. A sign attached to a second layer of fencing said VOLDEMORT INDUSTRIES in gothic script—a playful reference to the Harry Potter villain, known by the wizards in training as "He-Who-Must-Not-Be-Named." The only clue as to who actually owned this place was the picnic table with fixed seats, each painted a primary color:

red, blue, green, and yellow, familiar from Google's ubiquitous logo.

I had known when I contacted Google's media relations department that seeing the inside of a data center would be a long shot, given the notoriously tight lid the company kept on its facilities. But when I stressed that I wasn't interested in numbers (they change so fast anyway) but rather in the place itself—The Dalles and its character—they agreed to a visit. Certainly Google's presence in The Dalles was no longer a secret. There may not have been a sign outside (except for Voldemort Industries), but the company had joined the local chamber of commerce, begun to participate in community activities, donated computers to schools, planted a garden just outside its high steel fence, and was planning a public Wi-Fi network for downtown. Granted, this all came at the end of several years of bad press, in which Google's data center was portrayed as a poorly hidden smog-belching factory—an image incongruous with the clean white pages, friendly demeanor, and immediate access we otherwise associate with Google. Company officials had been vocal about turning over a new leaf, releasing some statistics from their data centers around the world, and even a short video tour. They seemed to agree that hiding their data centers was no longer the best policy. So the farce that came next surprised me.

Inside the security building, a pair of guards sat in front of a bank of video monitors, wearing blue polo shirts embroidered with a sheriff's badge nestled into the first "o" in "Google." Three visiting "Googlers" had come in ahead of me and were waiting for their security clearance, which involved having their retinas scanned from a machine that looked like a set of sleek coin-operated binoculars.

"Employee number?" the guard asked as each one approached the counter. "Step up to the machine."

Then the scanner took over the conversation, a robotic woman's voice, like a spaceship in a sci-fi movie. "Look into the mirror. Please stand closer." Snap. "Eye scan complete. Thank you." The visiting Googlers all giggled. Then the guard issued them a warning: make sure to scan in and scan out when you enter and exit the data center, because if the computer thinks you're still on the data center floor, it won't let you back in.

I wouldn't have that problem because—just as Young suspected—not only would I not see the data center floor, I wouldn't enter any buildings except the lunchroom. I began to realize I had come to The Dalles for a tour of the parking lot. Google's first rule of data center PR was: don't go in the data center.

I was greeted by a small entourage: Josh Betts, one of the facility's managers; an administrative assistant, Katy Bowman, who had spearheaded the community outreach; and a media handler who had driven in from Portland. Things were awkward from the beginning. When I pulled out my tape recorder, the media person leaned in for a close look at it, checking to make sure it wasn't a camera. With stiff smiles, we headed out into the rain, buzzed through a gate in the fence, and set out on foot across the campus. It felt like the back of a shopping mall, with broad parking lots, loading docks, and a tiny polite nod toward landscaping. The week before, I'd spoken with Dave Karlson, who managed the data center but was going to be away on vacation. "At the facility we'll hopefully be able to show you around what it looks and feels like to be at a Google data center," he said. But it took only a few moments of silence to make me realize that there'd be no guiding going on in this tour. Can you tell me about what we're looking at, and what these buildings do? I ventured. Betts carefully avoided looking at the media handler, pursed his lips, and stared at the pavement in front of him—an information void like a hung web page. I tried to make my question more specific. What about this building over here—a yellow warehouse-looking thing with steam drifting from a vent? Was it mostly for storage? Did it contain the computers that crawl the web for the search index? Did it process search queries? There were nervous glances back and forth. The handler skipped a couple steps to get within earshot.

"You mean what The Dalles does?" Betts finally responded. "That's not something that we probably discuss. But I'm sure that data is available internally."

It was a scripted nonanswer, however awkwardly expressed. Of course he knows what these buildings do—he manages the facility. He just wasn't about to tell me. But the march across the parking lot was an invitation to describe what I saw: There were two main data center buildings, each the shape and character of a distribution warehouse on the side of the highway. They were set on either end of an empty lot the size of a soccer field. There were two components to each of the buildings—a long low section and a taller end piece—that together formed an "L" kicking to the sky. On top of the L were the cooling towers, which gave off a hefty steam that rolled across the length of the building, like a Santa Claus beard. There were loading docks all around, but no windows. The roofs were clean. At the rear of each building was a series of generators encased in steel cabinets and attached with thick umbilicals of cables. Up close, the buildings' particularly unpleasant shade of beige-yellow could only have been chosen for being nondescript, on sale, or to make the place look like a penitentiary. Their signage—numbers only, no names—was perfectly painted and rational, with big, easy-to-read blue letters on beige. The roads had clean sidewalks with gravel neatly landscaped to their edges. Enormous lampposts spiked the campus, each topped by a halo of silver ball lights. The empty field—soon to house a third building—was filled with pickup trucks and modular construction offices. Just behind the high fence the Columbia River flowed steadily by.

As we approached the far edge of the property, Betts called security from his cell phone, and a few moments later a guard in a gray pickup pulled up. He unlocked a pedestrian gate, and we stepped through. We all admired the small garden tilled by Googlers in their off time, although it was too early in the season to see much growth. Beside the garden was another orange-and-white plastic pole, marking the location of the Q-Life fiber underground. Then we turned around and retraced our steps.

Near the entrance to Columbia House, which housed the dining room, Betts spoke up. "So you can see for yourself the campus. We've walked the perimeter. You can see what we have to deal with and what we've got going. In terms of the future you can get a sense of it just by taking a look around." I felt as if we were playing a puzzle game—perhaps the kind Google gives its employees as job applications. What was being left unsaid? Were they speaking in code? What was I supposed to be seeing?

A bearded man pedaled by on a sky-blue cruiser bike. "I think we're ready to go in!" my handler said.

Lunch was delicious. I ate organic salmon, a mixed green salad, and a peanut butter pudding for dessert. A handful of Googlers had been invited to join us, and as they sat down, my handler prompted each of them to say a few words about how much they liked living in The Dalles and how much they liked working at Google. "Can you tell Andrew what you like about working at Google and living in The Dalles?" she asked. Betts is a data center expert, second in charge of what I can only assume was among the most innovative facilities in the world, a key component of perhaps the greatest computing platform ever created. But he was sullen—preferring to say nothing at all than to risk stepping outside the narrow box PR had inscribed for him. We talked about the weather.

. I considered expressing my frustration at the kabuki going on. Wasn't Google's mission to make information available? Aren't you the best and the brightest, and eager to share what you all know? But I decided the silence wasn't their choice. It was bigger than them. Calling them out for it would have been unfair. Emboldened by my peanut butter cup, I eventually said only that I was disappointed not to have the opportunity to go inside a data center. I would have liked to have seen it. My handler's response was immediate: "Senators and governors have been disappointed too!" A guy came off the lunch line wearing a T-shirt that said PEOPLE WHO THINK THEY KNOW EVERYTHING ARE ANNOYING TO THOSE OF US WHO ACTUALLY DO.

It wasn't until I drove away that it began to sink in how strange the whole visit had been, easily the strangest visit I'd had to the

Internet. I hadn't learned anything from Google—except all the things I couldn't know. I wondered if I was being unfair, if the Orwellian atmosphere was just the side effect of Google's legitimate prerogative to maintain corporate secrets, and to protect our privacy. On its corporate website, I'd even read a little note about it (which later disappeared): "We realize that data centers can seem like 'black boxes' for many people, but there are good reasons why we don't reveal every detail of what goes on at our facilities, or where every data center is located," it said. "For one thing, we invest a lot of resources into making our data centers the fastest and most efficient in the world, and we're keen to protect that investment. But even more important is the security and privacy of the information our users place in our trust. Keeping our users' data safe and private is our top priority and a big responsibility, especially since you can switch to one of our competitors' products at the click of a mouse. That's why we use the very best technology available to make sure our data centers and our services remain secure at all times."

Google's famed mission statement is "to organize the world's information and make it universally accessible and useful." Yet at The Dalles, they'd gone so far as to scrub the satellite image of the data center on Google Maps—the picture wasn't merely outdated, but actively obscured. In dozens of visits to the places of the Internet, people I'd met had been eager to communicate that the Internet wasn't a shadowy realm but a surprisingly open one, dependent to its core on cooperation, on information. Driven by profit, of course, but with a sense of accountability. Google was the outlier. I was welcomed inside the gates, but only in the most superficial way. The not-so-subliminal message was that I, and by extension you, can't be trusted to understand what goes on inside its factory—the space in which we, ostensibly, have entrusted the company with our questions, letters, even ideas. The primary colors and childlike playfulness no longer seemed friendly—they made me feel like a schoolkid. This was the company that arguably knows the most about us, but it was being the most secretive about itself.

On my way into The Dalles, I'd stopped at the Bonneville Dam, the massive power plant—built and still operated by the Army Corps of Engineers—that spans the Columbia River. It was a fortress. Coming off the highway, I passed through a short tunnel, with a huge iron gate. An armed guard greeted me, asked if I was carrying any firearms, and searched my trunk. Then she tipped her hat and welcomed me in. There was a big visitor's center with a gift shop, exhibits about the construction of the dam and the ecology of the river, and a glass-walled room where you could see salmon swimming up the "fish ladder," on their way upstream to spawn. It was a classic American roadside attraction, a somewhat kitschy blend of big government and big landscape, all tied up in a complicated story of technological triumph and environmental tragedy. For an infrastructure geek—much less anyone who likes watching fish—the dam was a great stop.

I couldn't help but contrast the dam with the data center. One is owned by the government, the other by a public corporation; both are proud examples of American engineering. And they are functionally intertwined: it's the Bonneville Power Administration that, in part, drew Google to the region. But where the dam was welcoming, the data center was forbidding. What if Google opened a visitor's center, with a gift shop and a viewing gallery of all its servers? I think it would be a popular tourist attraction, a place to learn about what goes on behind Google's white screen. For now, though, the stance is the opposite. The data center is locked down, obscured.

While visiting the Internet I often felt like a pioneer, a first tourist. But the dam made me realize that might be temporary, that there'd be others behind me. So much of ourselves is in these buildings that Google's position will be tough to retain. I had gone to the Internet to see what I might learn from the visit. From Google, I hadn't learned much. Driving away, I preferred not to think about what Google knew about me.

There was another way of doing it. Google wasn't the only giant in the region. To the north was Quincy, where Microsoft, Yahoo!, Ask.com, and others all have major data centers. And just over a hundred miles due south from The Dalles was the town of Prineville, which might be as equally hard up as The Dalles but was truly off the beaten path. Yet Prineville was where Facebook had chosen to build its first ground-up data center, at a scale equal to The Dalles. That in itself struck me as an amazing testament to the advantages of central Oregon as a place to store data: four years after Google, Facebook had chosen it again.

Leaving The Dalles, a two-lane road rose up abruptly from the Columbia River Valley to the high plateau of central Oregon. Snow drifted across it in uneven streaks, crunching under my tires. There were actual tumbleweeds, blowing around like plastic bags in the city. And always there were the Bonneville Power Administration's high-tension power lines, which marched across the green sagebrush like columns of giant soldiers.

Prineville was a hundred-odd miles away. What was the significance of that distance? Why hadn't Facebook also moved into The Dalles? Or Quincy? On the map, they looked like neighbors. By virtue of being far away, and in a place where few people visit, it was easy to see it all as one place. But underneath the huge dome of sky, slowly lapping the miles, passing through little towns and empty crossroads, it was clear that each place had its own character, history, and people with a story to tell. The Internet "cloud," and even each piece of the cloud, was a real, specific place—an obvious reality that was only strange because of the instantaneity with which we constantly communicate with these places.

Just as I'd been lulled into the landscape's incredible expanse, I became distracted by the foreground, where spiked every few hundred yards into the red dirt of the highway shoulder were white plastic posts with orange caps. Eventually, I decided to pull over for a closer look at one. The cap read WARNING. BURIED FIBER OPTIC CABLE. NO DIGGING. LEVEL 3 COMMUNICATIONS.

Denver-based Level 3 operates one of the largest global backbone networks. One of its long-haul routes is here, in the dirt, most likely several hundred strands of fiber—although only a handful are likely "lit" with signals, while the remainder are "dark," awaiting future needs. Each individual fiber is capable of carrying terabits of data. But the bigness of that number (trillions!) and Level 3's geographic reach were the opposite of what really excited me. It was the freeze frame: the momentary presence of all that in this very particular spot. When you click on a little photo thumbnail and wait for the big image to load, those scattered pulses of light were passing beneath the edge of US 197, near the speck on the map labeled Maupin, Oregon—if only for a nearly infinitesimal fraction of a second. It is difficult to say for sure which roadside and which instant. But it is enough to remember that between here and there there's always a white-and-orange post on the side of a road. The path is continuous and true.

A few miles later I came over the top of a ridge and was rewarded with the Internet tourist's jackpot: a fiber-optic regeneration station, housing the equipment that amplified the light signals on their journey across the country. A barbed-wire fence enclosed an area the size of two tennis courts, with a gravel parking lot and three small buildings. When I got out for a closer look, the high desert wind whipped around and slammed the car door shut. Two of the buildings were steel walled, like shipping containers. The third was concrete and stucco, with multiple doors, like a self-storage locker. A zeppelin-shaped fuel tank the size of a sofa stood between them. On the fence was a white sign with red-and-black lettering: NO TRESPASSING. LEVEL 3. IN CASE OF EMERGENCY CALL. . . . Given that there was nothing around, none of the signal stopped here; all of it was merely received and re-sent by the racks of equipment inside, a necessary pit stop on the photons' journey through glass.

Directly across the road was an earlier generation's version of the same idea: an AT&T microwave site, hardened against nuclear attack, its big bunkerlike building, bigger than a house, looking sinister with its spindly antennae. The Level 3 encampment looked like the kind of sheds you'd find behind a gas station. I remembered the same contrast in Cornwall, where British Telecom's brutalist bunker stood in stark contrast with Global Crossing's more discreet house. I thought of the tilt-up concrete buildings in Ashburn, and the corrugated steel sheds in Amsterdam. The Internet had no master plan, and—aesthetically speaking—no master hand. There wasn't an Isambard Kingdom Brunel—the Victorian engineer of Paddington Station and the *Great Eastern* cable ship—thinking grandly about the way all the pieces fit together, and celebrating their technological accomplishment at every opportunity. On the Internet there were only the places in between, places like this, trying to disappear. The emphasis wasn't on the journey; the journey pretended not to exist. But obviously it did. I climbed up on the car for a better vantage point, doing what I could to climb up into that big sky. There was nobody around, the highway was empty, and there wasn't another house in sight. It was too windy to hear the hum of the machines.

Prineville was seventy-five more miles down the road. While The Dalles is in spitting distance of Portland, Prineville is tucked away in the middle of Oregon, far from the nearest interstate, in an area so remote that it was among America's last to be inhabited—until westward settlers noticed gold stuck in the hooves of lost cattle. Prineville is still a cowboy town, home to the Crooked River Roundup, a big annual rodeo. It's a place that has always fought to get its own, right down to founding a city-owned railroad after the main line passed it by—a nineteenth-century middle-mile network Nolan Young would appreciate. The City of Prineville Railway is still in operation ("Gateway to Central Oregon"), the last municipal-owned tracks in the entire country. But Prineville's biggest struggle came with the recent loss of its last major employer, Les Schwab Tire Centers—known for its "free beef" promotion—to Bend, a booming ski town with twelve Starbucks and a big Whole Foods, thirty miles away. Prineville fought to bring in Facebook, touting tax breaks that would save the company as much as $2.8 million a year, as a perk of its presence in an enterprise zone on the outskirts of town. But where Google insisted on secrecy, Facebook moved into Prineville with fanfare. Driving the town's main drag, with its intact 1950s roadside architecture, I spotted a WELCOME FACE-BOOK sign in a store window.

The data center sits on a butte above town, overlooking the Crooked River. Lining both sides of the wide road leading up to it are more white-and-orange fiber markers, like bread crumbs leading to the data center door. My first impression was of its overwhelming scale—long and low like a truck distribution center on the side of the highway. It was surprisingly beautiful, more visually assertive than most any piece of the Internet I'd seen, set on a shallow crest like a Greek temple. Where Google's buildings were aesthetically loose, with loading docks and appendages pointing every which way, Facebook's seemed tightly rational, a crisp human form in the sagebrush. It sits alone in the empty landscape, a clean concrete slab topped by a penthouse of corrugated steel. At the time I visited it was still under construction, with only the first large data center room finished. Three more phases extended out the back, like a caterpillar growing new body segments—the last one still showing the yellow insulating panels of endoskeleton. All together the four sections would total three hundred thousand square feet of space, the equivalent of a ten-story urban office tower. Construction on another building the same size would begin soon, and there was room on the property for a third. Across the country in Forest City, North Carolina, Facebook had begun construction on a sister building of the same design—which also happened to be fifty miles from Google's own massive data center in Lenoir, North Carolina.

Before I visited, I was predisposed to think of these big data centers as the worst kind of factories—black smudges on the virgin landscape. But arriving in Prineville I discovered what an industrial place it was already, from the vast hydroelectric infrastructure of the region to the remnant buildings of the timber industry that dotted the town. The notion of this data center despoiling the landscape was absurd. Prineville has long been a manufacturing town—and at the moment, what it needed most of all was more industry. What amazed me was that the data center had ended up here at all. This enormous building landed in the brush was an astonishing monument to the networked world. What's here is also in Virginia and Silicon Valley—that isn't surprising—but the logic of the network led this massive warehouse, this huge hard drive, to this particular town in Oregon.

I found Ken Patchett inside, leaning back in a brand-new Aeron chair with its tags still on. He sat at his desk in the sunny, open offices, his white iPhone earbuds stuck in his ears, finishing up a conference call. Before coming to Prineville to manage the data center, he'd held the same job in The Dalles, but it was hard to imagine him at Google. "My dog had more access to me there than my family!" he said. For all the Google-bots' silence, Patchett was uncensored. He's an enormous extrovert, with a booming voice and a winking sense of humor. At six feet, four inches, when he put his hard hat on for a walk through the building's unfinished sections, he looked like the iron workers still on the site. That fit: his job at Facebook wasn't about shaping information (at least not entirely), but the proper functioning of this huge machine.

Patchett grew up a military brat and then lived with his grandparents on their farm in New Mexico, milking cows before school. He wanted to be an iron worker or a policeman, but he dropped out of college when he couldn't play football anymore. He traveled the country for a job managing and servicing equipment at sawmills. "You want to talk about a wood chip, I am the guy to talk about a wood chip," he said. "And if your wood chips are no good, I can tell you how to make them better." The work even brought him to Prineville, where as a twenty-four-year-old he installed a chipper in the town mill for the current city manager. He had four kids and got into computers for the money. At a job fair in Seattle in 1998, he heard about a contract position at a Microsoft data center, paying $16 an hour. When he arrived at the place, he realized he'd help build it—one summer as an iron worker. "I walked in there and was, like, hey, I've been in here before! Then I saw this fellow lumber by and I thought he was a janitor, and next thing you know he comes out and it's the guy who's going to be interviewing me." He'd taken a couple technical classes, and the guy looked over his paperwork and asked, "What does that mean? Why should I hire you?" That taught him a lesson: "I don't know nothing. I know enough to figure my way around, but what I learned from taking all these classes is I don't know enough."

A month later, Microsoft had him running a data center.

"Like any good manager, I came in and painted the walls and brought in the plastic flowers." Then Microsoft moved him into the global networking team, and he celebrated the turn of the millennium sitting on the top of the AT&T building in Seattle with a satellite phone in his hand, in case the world ended. At Google a few years later, he started out managing The Dalles but soon was promoted out and ended up building data centers in Hong Kong, Malaysia, and China. He was in Beijing the day Google pulled out in 2010. "We left some boxes in there but they're not doing anything, just blinky lights," he assured me.

As we began talking about Facebook, I told Patchett how I was interested in why this building was here, of all places, seemingly in the middle of nowhere, but he cut me off. "Just because they don't have this one thing here, does that mean you forsake the whole community?" He shook his head. "Do you say screw 'em, let them eat cake?" It would be naive to think that Facebook came to Prineville to benefit the community—and indeed, Patchett came on board long after the site was chosen. But now that Facebook was there, he was determined for Facebook to be a part of Prineville. Facebook's ethos was bringing people together—perhaps, occasionally, more than they wanted to be. That extended to the data center. "We're not here to change the culture, just to integrate and be a part of this," Patchett offered.

To some extent this was a concerted PR effort to avoid repeating the mistakes that Google made in The Dalles, and the bad press that followed. Where Google had kept everything top secret, threatening legal action against anyone who even spoke its name, Facebook was determined to be wide open to this community. But it came wrapped in a broader statement about the openness of technology. At a press conference soon after I visited Prineville, Facebook launched the Open Compute project, where it shared the schematics of the entire data center, from the motherboard to the cooling system, and challenged others to use it as the starting point for improvement. "It's time to stop treating data centers like Fight Club," Facebook's director of infrastructure declared. But you could also look at the difference between Google and Facebook from the other angle: Facebook played fast and loose with our privacy while Google vehemently protected it. At the least, Patchett was happy to show off Facebook's data center. "Want to see how this shit really works?" he asked. "This has nothing to do with clouds. It has everything to do with being cold."

We started out in the glass-walled lobby filled with modern furniture in bright colors and photos of old-time Prinevillians on the wall. Facebook had hired an art consultant to poke around the town archives and choose images to decorate the place. (It made sense to me: If you're going to spend half a billion on hard drives, why not a few thousand on art?) Patchett leaned in close to one. "Look at the people—how'd you like to make her angry? And look at the hats. Everybody has their own hat," he said. "They all have their own style." He winked—that was a Facebook joke.

We passed the conference rooms named for local beers and entered a long, wide hallway with cavernous ceilings, like the stockroom of an IKEA. The overhead lights went on as we walked. Patchett swiped us through another doorway and into the first data center room, still in the process of being turned on. It was spacious and shiny, as big as a hotel ballroom, brand-new and uncluttered. On either side of an open central corridor were narrow aisles formed by high racks of black servers. In scale and shape, and with the concrete floors, the place felt like the underground stacks of a library. But in place of books were thousands of fluttering blue lights. Behind each light was a one-terabyte hard drive; the room contained tens of thousands of them; the building had three more rooms this size. It was the most data I'd ever seen in one place—the Grand Canyon of data.

And it was important stuff. This wasn't the dry database of a bank or government agency. Somewhere in here was stuff that was at least partly mine—among the most emotionally resonant bits around. But even knowing that, it still felt abstract. I knew Facebook as the thing on the screen, as a surprisingly rich medium for delivering personal news—of friends' new babies and jobs, health scares and vacations, first days of school and heart-wrenching memorials. But I couldn't avoid the breathtaking obviousness of what was physically in front of me: A room. Cold and empty. It all seemed so *mechanical*. What had I handed over to machines—these machines in particular?

"If you blew the 'cloud' away, you know what would be there?" Patchett asked. "*This*. This is the cloud. All of those buildings like this around the planet create the cloud. The cloud is a building. It works like a factory. Bits come in, they get massaged and put together in the right way, then packaged up and sent out. But everybody you see on this site has one job, that's to keep these servers right here alive at all times."

To minimize energy usage, the temperature in the data center is controlled with what amounts to a swamp cooler, rather than normal air conditioners. Cool outside air is let into the building through adjustable louvers near the roof; deionized water is sprayed into it; and fans push the conditioned air down onto the data center floor. "When the fans aren't on, and the air isn't being sucked through here—it's like a real cloud, dude," Patchett said. "I fogged this whole place up." Given Prineville's cold and dry climate, most of the year cooling is free. We stood beneath a broad hole in the ceiling, almost big enough to call an atrium. Daylight was visible along its upper edges. "If you stand right here and look up, you can see the fan bank," Patchett said. "The air hits this concrete floor and roils left and right. This whole building is like the Mississippi River. There's a huge amount of air coming in, but moving really slowly."

We left out the far end of the huge room and came into another wide hallway. "Here's my own personal storage room for stuff I don't really need," Patchett said. "And here's a bathroom I had no idea was back here until they put that sign on it." Behind another door was the second large data center space, a match to the one we'd just crossed but filled with server racks in various states of assembly. Behind this room would be two more like it—phases A, B, C, and D, ready for growth. Equipment had been arriving by the truckload every day. "We swarm on it like little server fairies, and by the morning, whhheeeee, there's all the blinky lights arranged in a nice order," Patchett said.

"But you've got to understand what your growth curve looks like. You want to make sure you don't overbuild. You want to be 10 percent ahead, although you're always 10 percent behind. But I'd rather be 10 percent behind than have a half a billion dollars of data center space sitting there." That reminded me that Patchett had the keys to Facebook's biggest single line item. The social network had recently raised a billion and a half dollars through a controversial private offering, orchestrated by Goldman Sachs. A significant fraction of that ended up in the back of a semitruck, chugging up this hill. But Patchett had been around long enough to be wary. "The Internet is a fickle thing," he said. "She's a crazy lady! So don't spend everything you've got up front because you may or may not use it. People get a swinging dick complex. Google built these monstrous data centers that are empty, you know why? Because it's fucking cool!"

After lunch, we climbed into Patchett's huge pickup truck and drove out on a dirt road through the woods behind the data center. Above us was the spur power line that Facebook had built off the main branch—a onetime expense that would pay for itself many times over. At a wide spot in the road, I could see the main lines running off into the northwest in the direction of The Dalles—and Portland, Seattle, Asia. At the edge of a bluff, we got out of the truck and looked out over the town of Prineville and toward the Ochoco Mountains. Immediately below us was an old sawmill, with a new power cogeneration plant built a decade ago but never used. "I think it's important to think about locally significant stuff when you're here," Patchett said. "What if when we're all grown up and ready to do something, what if we help get that back up and deliver twenty megawatts of power?" It wasn't a real plan, only a dream. In fact, Facebook had come under fire from Greenpeace for relying too heavily on coal power. But for Patchett, it was tied up in a broader vision about the future of data centers, and America. "If you lose rural America, you lose your infrastructure and your food. It's incumbent for us to wire everybody, not just urban America. The 20 percent of the people living on 80 percent of the land will be left behind. Without what rural America provides to urban America, urban America couldn't exist. And vice versa. We have this partnership." If in Oregon that was once about timber and beef, it now extended to data, of all things. The Internet was unevenly distributed. It wasn't everywhere at all—and the places where it wasn't suffered for it.

We climbed back in the truck and bounced back toward the data center, which emerged out of the woods like an oceanliner. Patchett was fiddling with his iPhone as he drove along the rutted road. "I just got an email," he said. Testing on the data center was done. "We are *live* on the Internet right now."

Epilogue

As everyone from Odysseus on down has pointed out, a journey is really only understood upon arriving home. But what did that mean when the place I was coming home from was everywhere?

The morning I left Oregon, I'd opened my laptop in the airport lounge to write some emails, read a few blog posts, and do the things I always do while sitting in front of the screen. Then, even more strangely, I did the same thing on the plane, paying the few bucks for the inflight Wi-Fi, flying above the earth but still connected to the grid. It was all one fluid expanse, the vast continent be damned—on the Internet's own terms, at least.

But I hadn't traveled tens of thousands of miles, crossed oceans and continents, to believe that was the whole story. This may not have been the most arduous of journeys—the Internet settles in mostly pleasant places—but it was a journey nonetheless. The science fiction writer Bruce Sterling voiced a popular sentiment when he wrote, "As long as I've got broadband, I'm perfectly at ease with the fact that my position on the planet's surface is arbitrary." But that ignores too much of the reality of how most of us live in the world. We're not merely connected, but rooted.

At some point soon after I arrived home from Oregon, I don't recall exactly when, I fished my laptop from my bag and opened it up. Then, silently, effortlessly, its hidden antenna latched on to the white wireless hub behind the couch, the one with the single green eye. That meant very little in logical terms, but it meant so much to me: I was home, back at my place on the network. When the squirrel nibbled on the cable a couple years earlier, I could watch him (or was it her?) from my desk in the small room I then used as an office. In the interim, the space had been given over to my daughter, her crib now occupying the same spot. The squirrel was still there. My daughter was big enough to stand up and look out the window and wave at him. Her corner of the world was a magical place, where animals told stories, baked cookies, and said good morning and good night. And it was a *small* place, a constrained geography; its specificity mattered. It mattered to me too.

It reminded me that while I'd seen many of the biggest monuments of the Internet, I hadn't answered one of the questions I had started out with: Where did the cable go from *here*? How did my piece of the Internet connect to the rest? On the sidewalk around the corner was a metal enclosure the size of a steamer trunk, which I suspected held the answer. A sign on it read CABLEVISION, my Internet provider. It was decorated with stickers advertising bands, and when I walked by late in the evenings—inevitably preoccupied with these words—I could hear it buzzing quietly.

But in a cruel irony, after so many of the Internet's doors had

opened, this one stayed mostly shut. Cablevision is a notoriously tight-lipped company, and only after months of phone calls
did I finally get a friendly engineer on the phone who sketched
the outlines of my cable's path. From the living room it passed
through a hole in the wall, down into my hundred-year-old
apartment building's basement, out into the backyard, past the
squirrel, across two neighbors' yards, and landed beneath the
steamer trunk in a thick bundle of cable—thicker, by far, than
the cables stretched across the ocean. Beside the steel trunk was
a manhole marked CATV. Inside of it was a fiber junction box, a
cylinder that looked a lot like a muffler, where all the cables from
the immediate neighborhood were aggregated into a few strands
of glass. On a street map this spot would be labeled CARLTON
AVE.; on Cablevision's network map, it was node 8M48, the *M*
denoting this area of north Brooklyn. The cable TV network was
first laid in the 1980s; since then, it had been steadily upgraded,
which in physical terms meant that fiber-optic cables reached out
closer and closer to customers' homes, expanding like the roots
of a tree, and with the capacity increasing each time. For now,
the fiber stopped at the curb; it would soon inevitably come all
the way to the door.

In the other direction, it went to the "head-end," a small,
industrial-looking building nearby surrounded by a fence, containing a piece of equipment known as a cable modem termina-
tion system, or CMTS. This was a special kind of router and
looked the part: a steel machine the size of a washer, sprouting
yellow wires, humming in a lonely room. All of Cablevision's
head-ends then plugged into just a few "master head-ends." The
one in Hicksville, Long Island—where Cablevision had its cor-

porate headquarters—was also the broadband Internet services center, or BISC, always pronounced like the soup. The large routers there were the same kind I'd seen at the PAIX in Palo Alto. They aggregated all the signals coming and going between Cablevision's customers and the rest of the Internet. And there, the trail got interesting.

Cablevision may not like to say much, but the company's network engineers can't help but be chattier. At a web page on the otherwise unused domain of cv.net, they maintained a list of the places where Cablevision connected to other networks. It looked familiar: 60 Hudson Street, 111 Eighth Avenue, Equinix Ashburn, Equinix Newark, Equinix Chicago, and Equinix Los Angeles. And since the logical routes were inherently visible, with a little extrapolation I could even get a sense for which networks Cablevision connected to: Level 3 (my favorite regen hut in Oregon), AT&T (I wonder if the company got a new sticker for its landing station mailbox), Hurricane Electric (with Martin Levy's slideshow of routers), and KPN (next door to the AMS-IX core). I knew that by virtue of being in New York I wasn't physically far from the center of the Internet, but it was striking to see how logically close I was.

I no longer saw the network as an amorphous blob, but as specific paths overlaid on the more familiar geography of the earth. The images in my head were precise: a short and familiar list of specific places. Admittedly, some of them were banal; I'd seen a lot of plain concrete buildings and linoleum-tiled, fluorescent-lit corridors. But just as many were beautiful—their beauty rooted in knowing the network's truths and the simple act of paying close attention to the world. To look for the Internet, I had gotten off the Internet. I had stepped away from my keyboard to look around and talk. No wonder then that some of the most vivid moments—the ones in which I felt most connected to these places—came outside the electronically locked doors. I remember in particular the evenings and—as any traveler would—the meals: the boardwalk fish shack in Costa da Caparica, Portugal, with the sun setting over the Atlantic (and the cable in its depths); the four-hundred-year-old country inn in Cornwall, where farmers in high rubber boots leaned on a stone fireplace; the brew pub in Oregon filled with skiers flipping through Facebook's blue-bordered screen on their phones (the data stored nearby).

But when I think of those moments I also think of being homesick—especially surrounded by people who were themselves home. I remember watching as Rui Carrilho, the Tata station manager in Portugal, greeted his wife and in-laws, who'd come to see the cable land on the beach (and the reason he'd missed dinner so many nights in a row). And I think of Jol Paling, who gave me a tour of Penzance's fishing docks after we dropped his son off at soccer practice—and who took a call on the way there from a colleague half a world away (and only halfway through his workday). Or Eddie Diaz, who, after spending all night underneath the Manhattan streets, headed home for a quick shower before going back out again for his wife's birthday. Or Ken Patchett setting down his giant mug of coffee to read the text message that arrived from his son—a sniper in the air force—who at that moment was sitting in a transport plane on the tarmac in Qatar. These guys aren't Steve Jobs or Mark Zuckerberg. They didn't invent anything, reshape any industries, or make a whole lot of money. They worked inside the global network and made it work. But they lived locally, as most of us do.

What I understood when I arrived home was that the Internet wasn't a physical world or a virtual world, but a human world. The Internet's physical infrastructure has many centers, but from a certain vantage point there is really only one: You. Me. The lowercase *i*. Wherever I am, and wherever you are.

Acknowledgments

When I began looking for the physical infrastructure of the Internet, I knew only in the vaguest terms how it might all fit together. From the very first moments, and throughout the entire process of researching and writing this book, I benefited from the extreme generosity of time and spirit on the part of many of the people who built and operate the networks that comprise the Internet. I've listed all of them in the chapter notes that follow. There are a few supreme experts whose contributions pervade the entire book, and to whom I'm especially grateful: Rob Seastrom, Eric Troyer, Anton Kapela, Martin Levy, Joe Provo, and Ilissa Miller. Internetworking is a complicated business that I've tried to make accessible and correct, but all errors, inaccuracies, or misunderstandings are entirely my own. Expert fact-checker Erik Malikowski helped avoid many.

My editor at *Wired* (now at Gizmodo), Joe Brown, saw the possibilities in this topic immediately and supported the early reporting that opened many crucial doors. My editor at *Metropo-*

lis, Martin Pedersen, encouraged me for years to get to work on a book already and gave me space and moral support when I finally did.

This book, and I, benefited immeasurably from a community of intellectual *makers*—writers, journalists, editors, teachers, filmmakers, curators—who are also all friends. I'm grateful to Tom Vanderbilt, Anthony Townsend, Ethan Youngerman, Kenny Salim, Beth Schwartzapfel, Stu Schwartzapfel, Mark Lamster, Astra Taylor, Kazys Varnelis, Tony Dokoupil, Alexis Madrigal, David Moldawer, Greg Lindsay, Sarah Fan, James Sanders, Jason Hutt, Paul Goldberger, David Schwartz, James Biber, Rupal Sanghvi, John Cary, Kenny Caldwell, Rosalie Genevro, Anne Rieselbach, Cassim Shepherd, Varick Shute, Greg Wessner, Nick Anderson, Seth Fletcher, Geoff Manaugh, Nicola Twilley, Ted Relph, Kanishka Goonewardena, Kirsten Valentine Cadieux, Nik Luka, Zack Taylor, and Laura Taylor.

I cannot imagine that there exists a more professional and talented team than the one that transformed an idea, and then a manuscript, into this book. My agent, Zoë Pagnamenta, was always two steps ahead: she made this happen. The manuscript benefited from more than its fair share of expert editing, with people looking out for it on both sides of the Atlantic. Jim Gifford at HarperCollins Canada enthusiastically gave it a home in my second home. Will Hammond at Viking in London offered subtle and profound suggestions the whole way through. At Ecco, Dan Halpern, Shanna Milkey, Rachel Bressler, Allison Saltzman, and Michael McKenzie together put out amazing and beautiful books; I'm grateful for the support and attention they've given this one. Hilary Redmon generously and enthusiastically adopted the project. My editor, Matt Weiland, pointed out the most scenic paths, unfailingly improved each draft, and brought an intensity to the production process that every book should be lucky enough to have.

And every writer should be lucky enough to receive the kind of support from the Blum and Pardo families that I did; not only were they always there cheering but—to a woman—they also offered many astute suggestions. I'm especially grateful to my parents for not once in twenty years suggesting that being a writer wasn't a noble, practical, and worthwhile profession. Phoebe was born with this book; so much of the thrill of writing it was the thought of her someday reading it. Above all, I'm grateful to Davina for her patience, love, and insight: my center.

Notes

Prologue

3 **In the F. Scott Fitzgerald story:** F. Scott Fitzgerald, *My Lost City: Personal Essays 1920–1940*, ed. James L. W. West III (Cambridge: Cambridge University Press, 2005), p. 115.

5 **Senator Ted Stevens of Alaska described the Internet:** The comments were made in the Senate Commerce, Science, and Transportation Committee's hearings for the "Communications, Consumers' Choice, and Broadband Deployment Act of 2006" on June 28, 2006. The full audio can be downloaded at http://www.publicknowledge.org/node/497.

5 **The *New York Times* fretted:** Ken Belson, "Senator's Slip of the Tongue Keeps on Truckin' Over the Web," *New York Times,* July 17, 2006 (http://www.nytimes.com/2006/07/17/business/media/17stevens.html).

6 **"The cyborg future is here":** Clive Thompson, "Your Outboard Brain Knows All," *Wired,* October 2007 (http://www.wired.com/techbiz/people/magazine/15-10/st_thompson).

7 **The Silicon Valley philosopher Kevin Kelly:** Kevin Kelly, "The Internet Mapping Project," June 1, 2009 (http://www.kk.org/ct2/2009/06/the-internet-mapping-project.php).

7 **Sure enough, one stepped forward:** Lic. Mara Vanina Oses "The Internet Mapping Project," June 3, 2009 (http://psiytecnologia.wordpress.com/2009/06/03/the-internet-mapping-project/).

10 **a "hard bottom," as Henry David Thoreau said of Walden Pond:** Henry David Thoreau, *Walden and Other Writings*, ed. Brooks Atkinson (New York: Modern Library, 1992).

1: The Map

My education in mapping the Internet, as well as many basics of the Internet's geography, came thanks to the patience of the excellent people at Tele-Geography, including Markus Krisetya, Alan Mauldin, Stephan Beckert, Bonnie Crouch, Roxanna Tran, Nicholas Browning, and former employee Bram Abramson—who also fished TeleGeography's first Internet report from the depths of his files and mailed it to me. In Milwaukee, Dave Janczak gave a great tour of the Kubin-Nicholson printing floor and filled in the company's history; Dr. Steven Reyer of the Milwaukee School of Engineering—and keeper of the "Milwaukee Architecture" website—rounded out some historical details about the building. I'm especially grateful to Jon Auer for opening up his particularly vivid piece of the Net. At the Oxford Internet Institute, Mark Graham helped my understanding of the challenges of mapping cyberspace.

12 **"Industrially, Milwaukee is known . . .":** *The WPA Guide to Wisconsin* (New York: Duell, Sloan and Pearce, 1941), pp. 247–48.

30 **In one well-known incident in February 2008:** For an excellent analysis of the outage, see the Renesys Corporation's report *The Day the YouTube Died,* June 2008 (http://www.renesys.com/tech/presentations/pdf/nanog43-hijack.pdf). For a song about the outage, see http://www.renesys.com/blog/2008/04/the-day-the-youtube-died-1.shtml.

2: A Network of Networks

A shelfful of books helped me to understand the Internet's history; I've listed them below. At UCLA, I'm grateful to Leonard Kleinrock who gave the better part of an afternoon to sharing stories. My understanding of the murky history of MAE-East was thanks to Steve Feldman, Bob Collet, and Rob Seastrom. On Tysons Corner, Paul Ceruzzi's book *Internet Alley* was indispensable. And out of nowhere, Matt Darling sent me the 1980 ARPANET directory, which he'd pulled out of the trash twenty years before.

Janet Abbate, Exploring the Internet (Cambridge: MIT Press, 1999).

Paul E. Ceruzzi, *Internet Alley* (Cambridge: MIT Press, 2008).

C. David Chaffee, *Building the Global Fiber Optics Superhighway* (New York: Kluwer Academic, 2001).

Katie Hafner and Michael Lyon, *Where Wizards Stay Up Late* (New York: Simon & Schuster, 1996).

Carl Malamud, *Exploring the Internet* (Englewood Cliffs, NJ: Prentice-Hall, 1993).

Stephan Segaller, *Nerds 2.0.1* (New York: T.V. Books, 1998).

Kazys Varnelis, *The Infrastructural City* (Barcelona, Spain: Actar, 2008).

35 **"the Internet lacks a central founding figure . . .":** Roy Rosenzweig, "Wizards, Bureaucrats, Warriors, and Hackers: Writing the History of the Internet," *American Historical Review* 103, no. 5 (December 1998): 1534.

35 **I should have known things wouldn't be so clear-cut:** Janet Abbate, *Exploring the Internet* (Cambridge: MIT Press, 1999), p. 2.

36 **"Not ideas about the thing but the thing itself":** Wallace Stevens, *The Collected Poems of Wallace Stevens* (New York: Alfred A. Knopf, 1954).

37 **As the philosopher Edward Casey writes:** Edward S. Casey, "How to Get from Space to Place in a Fairly Short Stretch of Time: Phenomenological Prolegomena," in *Senses of Place,* eds. Steven Feld and Keith H. Basso (Santa Fe: School of American Research, 1996).

38 **I tried to get a little postmodern kick:** Walter Kirn, *Up in the Air* (New York: Doubleday, 2001).

47 **"The Work of Art in the Age of Mechanical Reproduction":** As published in *Visual Culture: Critical Concepts in Media and Cultural Studies,* eds. Joanna Morra and Marquard Smith (New York: Routledge, 2006).

54 **Columbia law school professor Tim Wu points out:** Tim Wu, *The Master Switch* (New York: Alfred A. Knopf, 2010), p. 198.

54 **Winston Churchill said about architecture:** In a speech to the House of Commons, October 28, 1943, as quoted by the Churchill Center and Museum (http://www.winstonchurchill.org/learn/speeches/quotations).

63 **a claim repeated in James Bamford's bestselling 2008 book:** James Bamford, *The Shadow Factory* (New York: Anchor, 2008), p. 187.

63 **It's to this that Gore owed his purported claim:** For a good overview of the government's role in the development of the Internet, see *NSFNET: A*

Partnership for High-Speed Networking, Final Report 1987–1995, available at http://www.nsfnet-legacy.org/about.php.

65 Anthony Townsend has pointed out: Anthony M. Townsend, "Network Cities and the Global Structure of the Internet," *American Behavioral Scientist* 44, no. 10 (June 2001): 1697–1716.

3: Only Connect

From the earliest stages of this project, Eric Troyer at Equinix was a constant source of information and guidance; I couldn't have understood the Internet without his expertise. Also at Equinix, Aaron Klink, Dave Morgan, and Felix Reyes were generous with their time on both coasts; David Fonkalsrud at K/F Communications opened the door. And I'm grateful to Jay Adelson for the day he spent going down memory lane—and the faster ride, to the future, in his electric roadster.

70 As E. B. White said of New York: E. B. White, *Here Is New York* (New York: Little Bookroom, 2000).

70 the local venture capitalist John Doerr once described: Andy Serwer, "It Was My Party—and I Can Cry If I Want To," *Business 2.0,* March 2001.

70 As the MIT sociologist Sherry Turkle describes: Sherry Turkle, *Alone Together* (New York: Basic Books, 2011), pp. 155–56.

78 a "major global connectivity hub": Rich Miller, "Palo Alto Landlord Sues Equinix," *Data Center Knowledge,* September 20, 2010 (http://www.datacenterknowledge.com/archives/2010/09/20/palo-alto-landlord-sues-equinix/).

82 John Pedro would earn US patent 6,515,224 for his technique: Accessible at the United States Patent and Trademark Office's online database, http://patft.uspto.gov/.

83 the "Most Downloaded Woman": Guinness World Records, *Guinness World Records 2004* (New York: Guinness, 2003).

4: The Whole Internet

In the complicated and nuanced realm of Internet peering, I was grateful for the hours many people spent helping me to understand, in particular: Anton Kapela, Martin Levy of Hurricane Electric, Joe Provo, Ren Provo, Jim Cowie of Renesys, Jon Nistor, Josh Snowhorn, Daniel Golding of DH Capital, Sylvie LaPerrière, Michael Lucking, Rob Seastrom, Jay Hanke, Patrick Gilmore, and Steve Wilcox. In Frankfurt, several people opened the doors to their pieces of the Internet: Frank Orlowski and Arnold Nipper at DE-CIX; Martin Simon at Global Crossing; and Michael Boehlert at Anco-tel. Nikolaus Hirsch shared his crucial insights about the spirit of the city. In Amsterdam, Job Witteman, Henk Steenman, and Cara Mascini made sure I understood every part of AMS-IX's operation; Kees Neggers shared his knowledge about the history of the Internet in the Netherlands (and a great Indonesian restaurant); and Marc Gauw of NL-ix and Serge Radovic of Euro-IX brought a broader context to the world of Internet exchanges. Martin Brown was an enthusiastic fellow traveler on the walking tour of Amsterdam's data centers.

106 **As the website Wired.com explained:** Tony Long, "It's Just the 'internet' Now," *Wired News*, August 16, 2004 (http://www.wired.com/culture/ lifestyle/news/2004/08/64596).

106 **The writer Christine Smallwood:** Christine Smallwood, "What Does the Internet Look Like?," *Baffler* 2, no. 1 (2010): 8.

107 **"I wish," she concludes, "the Internet looked like Matt Damon:** Ibid., p. 12.

107 **On the episode of the television cartoon *South Park:*** *South Park,* "Over Logging," Season 12, Episode 6, originally aired on April 16, 2008.

108 **The British sitcom *The IT Crowd:*** *The IT Crowd*, "The Speech," Series 3, Episode 4, originally aired on December 12, 2008.

143 **extraordinary scene in Henry Adams's strange third-person autobiography:** Henry Adams, *The Education of Henry Adams* (Boston: Houghton Mifflin, 1918), p. 380.

146 **In his book *The Island at the Center of the World:*** Russell Shorto, *The Island at the Center of the World* (New York: Vintage, 2004), p. 28.

147 **"In the Netherlands, forts, canals, bridges, roads and ports . . .":** Jaap van Till, Felipe Rodriquez, and Erik Huizer, "Elektronische snel-weg moet hogere politieke prioriteit krijgen," *NRC Handelsblad,* August 21, 1997 (http://www.nrc.nl/W2/Nieuws/1997/08/21/Med/06 .html).

150 **a wonderful essay by the artist Robert Smithson:** Robert Smithson, "The Monuments of Passaic," *Artforum,* December 1967.

5: Cities of Light

At Brocade, Greg Hankins quickly arranged a visit, and Par Westesson was the ideal guide inside the machine. In New York, Ilissa Miller and Jaymie Scotto opened many doors, and I was especially grateful for their introduction to the inimitable Hunter Newby, who happily shared his insight and gave a great walking tour of Lower Manhattan. Michael Roark and Tesh Durvasula turned the lights on in some dark corners of the city's Internet. Victoria O'Kane and Ray La Chance happily accommodated my interest in seeing fiber-optic cables being laid, while Brian Seales and Eddie Diaz made it a fun night on the streets. John Gilbert at Rudin Management keeps the history alive at 32 Avenue of the Americas. In London, I'm grateful for the time and assistance of Tim Anker of the Colocation Exchange; Pat Vicary at Tata; John Souter, Jeremy Orbell, and Colin Silcock at the London Internet Exchange; Nigel and Benedicte Titley; Dionne Aiken, Michelle Reid, and Bob Harris at Telehouse; and Matthew Finnie and Mark Lewis at Interoute. James Tyler and Rob Coupland at Telecity spent the better part of a day showing off their impressive pieces of the Internet.

162 **In his essay, "Nature":** Stephen E. Whicher, *Selections from Ralph Waldo Emerson: An Organic Anthology* (Boston: Houghton Mifflin, 1957), p. 24.

163 **Google announced the purchase of 111 Eighth Avenue:** Rich Miller, "Google Confirms Purchase of 111 8th Avenue," *Data Center Knowledge* (http://www.datacenterknowledge.com/archives/2010/12/22/google-con firms-purchase-of-111-8th-avenue/).

164 **the "Ninth Avenue fiber highway":** The phrase caught on among real estate people and spread to the colocation providers. See, for example, Telx's website (http://www.telx.com/Facilities/telxs-new-york-city-colocation-a-interconnection-facility.html).

165 **a four-inch-diameter conduit will cost:** Empire City Subway's rates have not changed since 1987 (http://www.empirecitysubway.com/ratesbill.html).

172 **"Without a single hitch . . .":** "195 Broadway Deserted," *New York Times,* June 29, 1914.

181 **an urban landscape lifted from the pages:** J. G. Ballard, *High Rise* (London: Holt, Rinehart and Winston, 1977), p. 8.

186 **"Major colocation companies such as Telehouse . . .":** David Leppard, "Al-Qaeda Plot to Bring Down UK Internet," *Sunday Times,* March 11, 2007 (http://www.timesonline.co.uk/tol/news/uk/crime/article1496831.ece).

6: The Longest Tubes

In human terms, the world of undersea cables is an intimate one, and many people happily shared their knowledge and opened their facilities. At Global Crossing—now Level 3—Kate Rankin championed my interest to her colleagues, who collectively spent days answering my questions. In Rochester, Jim Watts, Mary Hughson, Louis LaPack, Mike Duell, and Nels Thompson provided the background info. Then in Cornwall, Jol Paling shared his beautiful part of the world. The Porthcurno Telegraph Museum is an invaluable resource for the history of undersea cables; archivist Alan Renton became a friend in the valley. At Hibernia Atlantic, Bjarni Thorvardarson welcomed me in and Tom Burfitt gave a great tour. At Tata Communications, Simon Cooper allowed everything to happen, most especially the chance to watch a cable land on the beach; his colleagues Janice Goveas, Paul Wilkinson, Rui Carrilho, and Anisha Sharma made it work. At TE Subcom, Courtney McDaniel arranged fascinating and hugely informative visits with Neal Bargano in Eatontown and Colin Young in Newington. Tom Standage's *The Victorian Internet* (New York: Walker, 2007) and John Steele Gordon's *A Thread Across the Ocean* (New York: Walker, 2002) filled in the fascinating history, while Richard Elliott at Apollo brought it up to the present. And every word written about this topic owes a debt to Neal Stephenson's epic 1996 article for *Wired,* "Mother Earth Mother Board" (available at http://www.wired.com/wired/archive/4.12/ffglass.html).

194 **In their continental scale, they invoked:** F. Scott Fitzgerald, *The Great Gatsby* (New York: Charles Scribner's Sons, 1953), p. 159.

195 **In 2004, Tata paid $130 million:** Ken Belson, "Tyco to Sell Undersea Cable Unit to an Indian Telecom Company," *New York Times,* November 2, 2004 (http://www.nytimes.com/2004/11/02/business/02tyco.html).

195 **CEO Dennis Kozlowski—who was convicted of grand larceny:** "Top Ten Crooked CEOs," *Time,* June 9, 2009 (http://www.time.com/time/specials/packages/article/0,28804,1903155_1903156_1903152,00.html).

206 **The engineer's report of the "Porthcurnow—Gibraltar No. 4 Cable":** Available at the archives of the Porthcurno Telegraph Museum.

7: Where Data Sleeps

Any journalist who has waded into the world of data centers quickly learns that the gold standard for coverage is Rich Miller's blog, *Data Center Knowledge*. His posts were a constant source of news and context, and I'm grateful for his friendly support of this project. At Facebook, Ken Patchett told me everything

I wanted to know and more, while Lee Weinstein kept him talking. In The Dalles, Nolan Young told a straight story, on short notice. At Google, I'm grateful to Kate Hurowitz for opening the door (if only slightly), and to Dave Karlson, Katy Bowman, Josh Betts, and Marta George for their time. Michael Manos—employed by AOL when we spoke—is an unmatched participant and observer of the industry. In Virginia, Dave Robey at QTS and Norm Laudermilch at Terremark happily shared their respectively monstrous data centers, with welcome assistance from their colleagues Kevin O'Neill and Xavier Gonzalez.

230 **2 percent of the world's electricity usage:** Greenpeace International, "How Dirty Is Your Data?," April 21, 2011 (http://www.greenpeace.org/international/en/publications/reports/How-dirty-is-your-data/).

231 **In 2011, Facebook reported:** Facebook engineer Justin Mitchell provided the number on the website Quora, January 25, 2011 (http://www.quora.com/How-many-photos-are-uploaded-to-Facebook-each-day).

231 **Google confirms at least one billion searches per day:** Matt McGee, "By The Numbers: Twitter Vs. Facebook Vs. Google Buzz," *Search Engine Land,* February 23, 2010 (http://searchengineland.com/by-the-numbers-twitter-vs-facebook-vs-google-buzz-36709).

236 **Its total cost was $1.8 million:** For an account of Google's arrival in The Dalles, see Steven Levy, *In the Plex* (New York: Simon & Schuster, 2011), pp. 192–95.

238 **"It was visionary—this little town . . .":** Ibid., p. 192.

248 **I'd even read a little note about it:** The site has since been changed, but it was accessible as of June 2011 at http://www.google.com/corporate/datacenter/index.html; a copy is preserved here: http://kalanaonline.blogspot.com/2011/02/where-is-your-data-google-and-microsoft.html.

257 **Where Google had kept everything top secret:** John Markoff and Saul Hansell, "Hiding in Plain Sight, Google Seeks More Power," *New York Times,* June 14, 2006 (http://www.nytimes.com/2006/06/14/technology/14search.html).

258 **"It's time to stop treating data centers like Fight Club":** Maggie Shiels, "Facebook Shares Green Data Centre Technology," *BBC News,* April 8, 2011 (http://www.bbc.co.uk/news/technology-13010766).

261 **Facebook had come under fire from Greenpeace:** Elizabeth Weingarten, "Friends Without Benefits," *Slate,* March 7, 2011 (http://www.slate.com/id/2287548/).

Index

Index

Index

Index

Index

Index

ANDREW BLUM is a journalist writing about infrastructure, technology, architecture, and design. He is the author of *The Weather Machine: A Journey Inside the Forecast*, and his magazine writing has appeared in numerous publications, including *Wired*, *Popular Science*, *Vanity Fair*, and the *New York Times*. He lives with his family in Brooklyn, New York.